NANO AND BIO HEAT TRANSFER AND FLUID FLOW

T0311903

NANO AND BIO HEAT TRANSFER AND FLUID FLOW

PROFESSOR MAJID GHASSEMI
Mechanical Faculty
K.N. Tossi University of Technology

DR. AZADEH SHAHIDIAN
Mechanical Faculty
K. N. Toosi University of Technology

ACADEMIC PRESS

An imprint of Elsevier
elsevier.com

Academic Press is an imprint of Elsevier
125 London Wall, London EC2Y 5AS, United Kingdom
525 B Street, Suite 1800, San Diego, CA 92101-4495, United States
50 Hampshire Street, 5th Floor, Cambridge, MA 02139, United States
The Boulevard, Langford Lane, Kidlington, Oxford OX5 1GB, United Kingdom

Notices
Knowledge and best practice in this field are constantly changing. As new research and experience broaden our understanding, changes in research methods, professional practices, or medical treatment may become necessary.

Practitioners and researchers must always rely on their own experience and knowledge in evaluating and using any information, methods, compounds, or experiments described herein. In using such information or methods they should be mindful of their own safety and the safety of others, including parties for whom they have a professional responsibility.

To the fullest extent of the law, neither the Publisher nor the authors, contributors, or editors, assume any liability for any injury and/or damage to persons or property as a matter of products liability, negligence or otherwise, or from any use or operation of any methods, products, instructions, or ideas contained in the material herein.

Library of Congress Cataloging-in-Publication Data
A catalog record for this book is available from the Library of Congress

British Library Cataloguing-in-Publication Data
A catalogue record for this book is available from the British Library

ISBN: 978-0-12-803779-9

For information on all Academic Press publications visit our website at
https://www.elsevier.com/books-and-journals

Working together
to grow libraries in
developing countries

www.elsevier.com • www.bookaid.org

Publisher: Joe Hayton
Acquisition Editor: Fiona Geraghty
Editorial Project Manager: Ashlie Jackman
Production Project Manager: Susan Li
Designer: Matthew Limbert

Typeset by TNQ Books and Journals

To my daughter, Fatimah, son, Alireza, and my wife, Monir
Majid

To my mother, Mansoureh, my father, Hossein, and my husband Hossein
Azadeh

CONTENTS

AUTHORS' BIOGRAPHY

Professor Majid Ghassemi

Majid Ghassemi is a Professor in the Mechanical Engineering Department at the K. N. Toosi University of Technology, one of the most prestigious technical universities in Tehran, Iran. He is currently a Visiting Professor at the Centre for Hydrogen and Fuel Cell Research at the University of Birmingham. Professor Ghassemi has been endorsed by The Royal Academy of Engineering of the United Kingdom as an Exceptional Talent. The United Kingdom selects only a few scientists per year from all over the world as an Exceptional Talent.

Professor Ghassemi teaches graduate and undergraduate courses and conducts research in the area of heat transfer and its application in bio- and microsensors, drug delivery, fuel cells, microchannels, and alternative energy. He has over 20 years of academic and industrial experience and served as the President of the K. N. Toosi University of Technology from 2010 through 2013. Professor Ghassemi has also served in several public and private boards and panels, supervised several undergraduate, masters, and PhD students, and published several books and many journal and conference papers.

He is currently the Editor-in-Chief of the *International Journal of Prevention and Treatment* and Managing Editor of the *American Journal of Mechanical Engineering* (AJME) as well as Editorial Board member for many international journals. He also serves as member in several international conferences. Professor Ghassemi received his PhD in Mechanical Engineering from Iowa State University in 1993.

Dr. Azadeh Shahidian

Azadeh Shahidian is an Assistant Professor in the Mechanical Engineering Department at K. N. Toosi University of Technology in Tehran, Iran. She received her BSc and PhD degree in Mechanical Engineering from K. N. Toosi University of Technology in Tehran, Iran. Dr. Shahidian's research focused on the numerical and experimental investigation of nano and biofluid properties and behavior. She currently teaches graduate and undergraduate courses and conducts research in the area of heat transfer and its application in drug delivery, energy management, and biofluid mechanics. Dr. Shahidian has 15 years of academic and industrial experience and has supervised several undergraduate, masters, and PhD students. Dr. Shahidian has published over 25 papers in well-known journals and conferences.

PREFACE

Progress in the field of science and engineering has significantly improved the human quality of life. Among other technologies, nano-biotechnology is one of the key technologies of the 21st century that has the potential to significantly improve the quality of life of patients. Convergences of nanotechnology have created several new interdisciplines such as nanochemistry, nanomedicine, and nanomaterials. The convergence of nanotechnology and biotechnology established nano-biotechnology disciplines, the science at the nanometer length scales. Nano-biotechnology is concerned with medical treatment using engineering tools (i.e., drug delivery, preparation of nanodrugs, and various methods of targeted transport of active substances). The nanotherapy vision of the future is treatment of patients with individually tailored medicines ("personalized medicine") at the molecular level as soon as the disease is in the development stage.

Nanosize particles and their application in engineering and science have been around for several decades. However nano-biotechnology, as a modern interdisciplinary science, started from the 19th century and developed rapidly. Nanotechnology was first introduced in the 1950s by Richard P. Feynman. There is hardly any area of medicine that could not benefit from the prospects nano-biotechnology offers. Nano-biotechnologies open up new possibilities above all in the field of regenerative medicine. In the coming decade nano-biotechnology applications will gain importance in medicine and medical technology. Nevertheless, the new possibilities also involve risks and raise sociological and ethical questions, which must be analyzed and debated.

The purpose of this book is to investigate the use of nanoparticles for bioapplication from an engineering perspective. The book introduces the mechanisms underlying thermal and fluid interaction of nanoparticles with biological systems. The key point that distinguishes this book from other similar works is practical approach. This book will help readers translate theory into real-world applications, such as drug delivery and lab-on-a-chip. The book also discusses the complications that can arise in a biological system at the nanoscale level. This book is ideal for students and early career researchers, engineers conducting experimental work on relevant applications, or those who develop computer models to investigate/design these systems. Although this book primarily addresses the scientific community, it offers valuable and accessible information for a wide range of readers interested in the current technological progress with direct relevance to the biomedical field. Content coverage includes biofluid mechanics, transport phenomena, micro/nano fluid flows, and heat transfer.

ACKNOWLEDGMENT

We, the authors, would like to express our greatest respect to our parents *without whom none of this would be possible*. We would also like to extend our highest gratitude to our immediate family for providing endless support and encouragement during the entire endeavor. Our gratitude goes to Dr. Sogand Noroozizadeh for her assistance with editing some of the book chapters. We also appreciate assistance from our master and PhD students at the Nano and Fuel Cell Laboratory in the K.N. Toosi University of Technology. Our special thanks go to Dr. Hossein Afshar, Dr. Mohammad Reza Habibi, Dr. Negin Maftouni, Mr. Roozbeh Ayani, Seyed Mohammad Ali Nemati, Majid Kamvar, and Mohammad Reza Mohammadi, and Mrs. Zahra Abbasi.

We also appreciate the Elsevier staff, Ashlie M. Jackman, Maria Convey, and Fiona Geraghty, for their support and assistance throughout the entire process. We also appreciate the referee's time and useful suggestions.

Last and but not least we beg forgiveness of all those who have been with us over the course of the book and whose names we have failed to mention.

CHAPTER 1

History of Bio-Nano Fluid Flow

Recent application of nanotechnology in biomechanic concept, such as medicine, drug delivery, and cancer therapy, has generated a lot of interest in investigation of nano bio heat transfer and fluid flow. According to the applications of mechanical engineering and nanotechnology in medicine, employing the laws of fluid mechanics and heat transfer in nano–bio fluid application seems to be essential. Before introducing fluid mechanics and heat transfer applications to nano bio systems, a brief history of book topics, fluid mechanics, heat transfer, biomedical engineering, and nanotechnology are reviewed in this section.

1.1 FLUID MECHANICS AND HEAT TRANSFER, OLDER SCIENCES

1.1.1 Fluid Mechanics

Fluid mechanics is the branch of physics that deals with the mechanics of fluids (liquids, gases, and plasmas) and the forces on them. A fluid is a substance that cannot resist a shear stress by a static deflection and deforms continuously as long as the shear stress is applied. Fluid mechanics can be divided into fluid statics or the study of fluids at rest; and fluid dynamics or the study of the effect of forces on fluid motion. Fluid mechanics has a wide range of applications, including mechanical engineering, chemical engineering, geophysics, astrophysics, and biology. Fluid mechanics, especially fluid dynamics, is an active field of research with many problems that are partly or wholly unsolved. Commercial code based on numerical methods is used to solve the problems of fluid mechanics. The principles of these methods are developed by computational fluid dynamics (CFD), a modern discipline that is devoted to this approach to solve the aforementioned problems.

A short history of fluid mechanics from the beginning up to now is as follows:

- The fundamental principles of hydrostatics and dynamics were given by Archimedes (285–212 BC) in his work On Floating Bodies, around 250 BC. Archimedes has developed the law of buoyancy, also known as Archimedes' Principle [1].
- Islamicate scientists, particularly Abu Rayhan Biruni (973–1048) and later Al-Khazini (1115–30), were the first to apply experimental scientific methods to fluid mechanics, especially in the field of fluid statics, to determine specific weights.
- In the 9th century, Banū Mūsā brothers' *Book of Ingenious Devices* described a number of early automatic controls in fluid mechanics [2]. Two-step level control for fluids, which was an early form of discontinuous variable structure controls, was developed by the Banū Mūsā brothers [3]. They also described an early feedback controller for fluids.

Nano and Bio Heat Transfer and Fluid Flow
ISBN 978-0-12-803779-9, http://dx.doi.org/10.1016/B978-0-12-803779-9.00001-7

- In 1206, Al-Jazari's *Book of Knowledge of Ingenious Mechanical Devices* described many hydraulic machines [4]. Of particular importance were his water-raising pumps. Al-Jazari also invented a twin-cylinder reciprocating piston suction pump, which included the first suction pipes, suction pumping, and double-action pumping, and made early uses of valves and a crankshaft-connecting rod mechanism.
- Then, Leonardo da Vinci (1452—1519) derived the equation of conservation of mass in a one-dimensional steady flow [5].
- A Frenchman, Edme Mariotte (1620—84), built the first wind tunnel and tested models in it.
- The effects of friction and viscosity in diminishing the velocity of running water were noticed in the *Principia* of Sir Isaac Newton, which threw much light upon several branches of hydromechanics.
- In 1687, Isaac Newton (1642—1727) postulated his laws of motion and the law of viscosity of the linear fluids, which is now called newtonian. The theory first yielded to the assumption of a "perfect" or frictionless fluid, and 18th century mathematicians (Daniel Bernoulli, Leonhard Euler, Jean d'Alembert, Joseph-Louis Lagrange, and Pierre-Simon Laplace) introduced many beautiful solutions to frictionless-flow problems. At the end of the 19th century, unification between experimental hydraulics and theoretical hydrodynamics finally began.
- William Froude (1810—79) and his son Robert (1846—1924) developed laws of model testing.
- Lord Rayleigh (1842—1919) proposed the technique of dimensional analysis, and Osborne Reynolds (1842—1912) published the classic pipe experiment in 1883, which showed the importance of the dimensionless Reynolds number named after him.
- Meanwhile, viscous-flow theory was available but unexploited since Navier (1785—1836) and Stokes (1819—1903) had successfully added the newtonian viscous terms to the governing equations of motion. Unfortunately, the resulting Navier—Stokes equations were too difficult to analyze arbitrary flows.
- In 1904 a German engineer, Ludwig Prandtl (1875—1953), showed that fluid flows with small viscosity (water and air flows) can be divided into a thin viscous layer, or boundary layer, near solid surfaces and interfaces, patched onto a nearly inviscid outer layer, where the Euler and Bernoulli equations apply. Boundary layer theory is the most important tool in modern flow analysis.

1.1.2 Heat Transfer

In physics, heat is defined as the transfer of thermal energy across a well-defined boundary around a thermodynamic system. Heat transfer is a process function (or path function). It means that the amount of heat transferred that changes the state of a system depends not only on the net difference between the initial and final states of the process but also on how that process occurs. The rate of heat transfer is dependent on the temperatures of the systems and the properties of the intervening medium through which the heat is transferred. In engineering contexts, the term heat is taken as synonymous to thermal energy. This usage has its origin in the historical interpretation of heat as a fluid (caloric) that can be transferred by various causes. The transport equations for thermal energy (Fourier's law), mechanical momentum (Newton's law for fluids), and mass transfer (Fick's laws of diffusion) are similar, and analogies among these three transport processes have been developed to facilitate prediction of conversion from any one to the others. The fundamental modes of heat transfer are Advection, Conduction or diffusion, Convection, and Radiation. Types of phase transition occurring in the three fundamental states of matter include: deposition, freezing and solid to solid transformation in solid, boiling/evaporation, recombination/deionization, and sublimation in gas, and condensation and melting/fusion in liquid.

The history of heat has a prominent place in the history of science. The ancients viewed heat as related to fire. The ancient Egyptians in 3000 BC viewed heat as related to origin mythologies. It traces its origins to the first hominids to make fire and to speculate on its operation and meaning to modern day physicists who study the microscopic nature of heat. The history of heat is a precursor for developments and theories in the history of thermodynamics.

- Around 500 BC, the Greek philosopher Heraclitus and Hippocrates proposed the first theory about heat and its principles.
- In the 11th century AD, Abū Rayhān Bōrūnō cites movement and friction as causes of heat, which, in turn, produces the element of fire, and lack of movement as the cause of cold near the geographical poles [6].
- Around 1600, the English philosopher and scientist Francis Bacon surmised that heat itself, its essence and quiddity, is motion and nothing else.
- In 1761, Scottish chemist Joseph Black discovered that ice absorbs heat without changing the temperature when it is melting. Between 1759 and 1763, he formulated a theory of latent heat on which his scientific fame chiefly rests and also showed that different substances have different specific heats.
- James Watt invented the Watt engine. The ability to use heat transfer to perform work allowed the invention and development of the steam engine by inventors such as Thomas Newcomen and James Watt. In addition, in 1797 a cannon manufacturer Sir Benjamin Thompson, Count Rumford, demonstrated that through the use of friction it was possible to convert work to heat [7].
- In 1824 the French engineer Sadi Carnot set the importance of heat transfer: "production of motive power is due not to an actual consumption of caloric, but to its transportation from a warm body to a cold body, i.e., to its re-establishment of equilibrium." According to Carnot, this principle applies to any machine set in motion by heat [7].

- The work of Joule and Mayer demonstrated that heat and work were equivalent forms of energy, and led to the statement of the principle of the conservation of energy by Hermann von Helmholtz in 1847.
- In 1850, Clausius demonstrated that caloric theory could be reconciled with kinetic theory provided that the conservation of energy was employed rather than the movement of a substance and stated the First Law of Thermodynamics.

1.2 WHAT IS BIOENGINEERING?

The word bioengineering was coined by the British scientist and broadcaster Heinz Wolff in 1954 [8]. Biological engineering or bioengineering (including biological systems engineering) is the application of concepts and methods of biology to solve real-world problems related to life sciences, using analytical methods and simulation tools of engineers. Biological engineering employs knowledge and expertise from a number of pure and applied sciences such as mass and heat transfer, kinetics, biocatalysts, biomechanics, fluid mechanics, and thermodynamics. Bioengineering is used in the design of medical devices, diagnostic equipment, biocompatible materials, renewable bioenergy, ecological engineering, and agricultural engineering.

For example, biomimetics is a branch of biological engineering, which strives to find ways in which the structures and functions of living organisms can be used as models for the design and engineering of materials and machines. In nonmedical aspects, bioengineering is closely related to biotechnology, nanotechnology, and 3D printing. Physicist Richard Feynman theorized about the idea of a medical use for these biological machines that are introduced into the body to repair or detect damages and infections. The first biological engineering program was created at Mississippi State University in 1967, making it the first biological engineering curriculum in the United States [9]. More recent programs have been launched at the Massachusetts Institute of Technology (MIT) and Utah State University.

The word "biomechanics" (1899) and the related "biomechanical" (1856) refer to the study of the mechanical principles of living organisms, particularly their movement and structure [10]. Biomechanics is closely related to engineering, because it often uses traditional engineering sciences to analyze biological systems. Usually biological systems are much more complex than man-built systems. Numerical methods are, hence, applied to almost every biomechanical study. Research is done in an iterative process of hypothesis and verification, including several steps of modeling, computer simulation, and experimental measurements. Applied subfields of biomechanics include: Soft body dynamics, Kinesiology, Animal locomotion and Gait analysis, Musculoskeletal and Orthopedic biomechanics, Cardiovascular biomechanics, Ergonomy, Human factors engineering and Occupational biomechanics, Implant (medicine), Orthotics and Prosthesis, Rehabilitation, Sports biomechanics, Allometry, and Injury biomechanics.

Some simple examples of biomechanics research include the investigation of the forces that act on limbs, the aerodynamics of bird and insect flight, the hydrodynamics of swimming in fish, and locomotion, in general, across all forms of life, from individual cells to whole organisms. The biomechanics of human beings is a core part of kinesiology. As we develop a greater understanding of the physiological behavior of living tissues, researchers are able to advance the field of tissue engineering as well as develop improved treatments for a wide array of pathologies.

Biological fluid mechanics, or biofluid mechanics, is the study of both gas and liquid fluid flows in or around biological organisms. An often studied liquid biofluids problem is that of blood flow in the human cardiovascular system. Under certain mathematical circumstances, blood flow can be modeled by the Navier—Stokes equations. In vivo whole blood is assumed to be an incompressible Newtonian fluid. However, this assumption fails when considering forward flow within arterioles. At the microscopic scale, the effects of individual red blood cells become significant and whole blood can no longer be modeled as a continuum. When the diameter of the blood vessel is just slightly larger than the diameter of the red blood cell, the Fahraeus—Lindqvist effect occurs and there is a decrease in the wall shear stress [11]. However, as the diameter of the blood vessel decreases further, the red blood cells have to squeeze through the vessel and often can only pass in a single file. In this case, the inverse Fahraeus—Lindqvist effect occurs and the wall shear stress increases. An example of a gaseous biofluids problem is that of human respiration. Also, respiratory systems in insects have been studied for bioinspiration to design improved microfluidic devices. Over the past decade, the Finite element method has become an established alternative to in vivo surgical assessment. The main advantage of Computational Biomechanics lies in its ability to determine the endoanatomical response of an anatomy, without being subject to ethical restrictions.

Biological engineering is also called bioengineering by some colleges, and biomedical engineering is called bioengineering by others and it is a rapidly developing field with fluid categorization. The differentiation between biological engineering and biomedical engineering can be unclear as many universities loosely use the terms "bioengineering" and "biomedical engineering" interchangeably. Biomedical engineers specifically focus on applying biological and other sciences toward medical innovations, whereas biological engineers principally focus on applying engineering principles to biology, but not necessarily for medical uses.

1.3 HISTORY OF NANOTECHNOLOGY

Nanotechnology ("nanotech") is manipulation of matter on the atomic, molecular, and supramolecular scale. The description of nanotechnology referred to the particular technological goal of precise manipulation of atoms and molecules for fabrication of macroscale products [12]. A more generalized description of nanotechnology was subsequently established by the National Nanotechnology Initiative, which defines nanotechnology as the manipulation of matter with, at least, one dimension, whose size varies from 1 to 100 nm [13]. This definition reflects the fact that quantum mechanical effects are important at this quantum-realm scale, and so the definition shifted from a particular technological goal to a research category inclusive of all types of research and technologies that deal with the special properties of matter that occur below the given size threshold. Nanotechnology may be able to create many new materials and devices with a vast range of applications, such as in nanomedicine, nanoelectronics, biomaterials energy production, and consumer products.

The history of nanotechnology traces the development of the concepts and experimental work falling under the broad category of nanotechnology. Although nanotechnology is a relatively recent development in scientific research, the development of its central concepts happened over a longer period of time.

In 1959 the nanotechnology discussion was started by a renowned physicist Richard Feynman based on his talk, "There's Plenty of Room at the Bottom," in which he described the possibility of synthesis via direct manipulation of atoms. The term "nano-technology" was first used by Norio Taniguchi in 1974, although it was not widely known [14]. Inspired by Feynman's concepts, K. Eric Drexler used the term "nanotechnology" in his 1986 book *Engines of Creation: The Coming Era of Nanotechnology*, which proposed the idea of a nanoscale "assembler," which would be able to build a copy of itself and of other items of arbitrary complexity with atomic control [12]. After Feynman, scholars studying the historical development of nanotechnology have concluded that his actual role in catalyzing nanotechnology research was limited, based on recollections from many of the people who were active in the nascent field in the 1980s and 1990s. Chris Toumey, a cultural anthropologist at the University of South Carolina, found that the published versions of Feynman's talk had a negligible influence in the 20 years after it was first published and not much more influence in the decade after the Scanning Tunneling Microscope was invented in 1981 [15,16].

The Japanese scientist Norio Taniguchi of the Tokyo University of Science was the first to use the term "nano-technology" in a 1974 conference [14] to describe semiconductor processes, such as thin film deposition and ion beam milling characteristic control on the order of a nanometer. His definition was, "'Nano-technology' mainly consists of the processing of separation, consolidation, and deformation of materials by one atom or one molecule." However, the term was not used again until 1981 when Eric Drexler published his first paper on nanotechnology in 1986 [12]. His PhD work at the MIT Media Lab was the first doctoral degree on the topic of molecular nanotechnology and (after some editing) his thesis, "Molecular Machinery and Manufacturing with Applications to Computation," was published as *Nanosystems: Molecular Machinery, Manufacturing, and Computation* [13], which received the Association of American Publishers award for Best Computer Science Book of 1992.

The emergence of nanotechnology in the 1980s was caused by the convergence of experimental advances such as the invention of the scanning tunneling microscope in 1981 and the discovery of fullerenes in 1985, with the elucidation and popularization of a conceptual framework for the goals of nanotechnology beginning with the 1986 publication of the book *Engines of Creation*. The field was subject to growing public awareness and controversy in the early 2000s, with prominent debates about both its potential implications as well as the feasibility of the applications envisioned by advocates of molecular nanotechnology, and with governments moving to promote and fund research into nanotechnology. The early 2000s also saw the beginnings of commercial applications of nanotechnology, although these were limited to bulk applications of nanomaterials rather than the transformative applications envisioned by the field.

One of the useful findings of nanotechnology is nanofluid. Fluids with suspended nanoparticles (below 100 nm) are called nanofluids, a term first proposed by Choi in 1995 of the Argonne National Laboratory, USA [17]. The nanoparticles used in nanofluids are typically made of metals, oxides, carbides, or carbon nanotubes. Common base fluids include water, ethylene glycol, and oil. Nanofluid is considered to be the next-generation heat transfer fluids as they offer exciting new possibilities to enhance heat transfer performance compared with pure liquids. Researchers have demonstrated that nanofluids (such as water or ethylene glycol) with CuO or Al_2O_3 nanoparticles exhibit enhanced thermal conductivity [17]. Thus, the use of nanofluids, for example, in heat exchangers, may result in energy and cost savings and should facilitate the trend of device miniaturization. More exotic applications of nanofluids can be envisioned in biomedical engineering and medicine in terms of optimal nano-drug targeting and implantable nano-therapeutic devices [18].

The key point that distinguishes this book from other similar works is the fact that it has a practical point of view in some heat transfer and fluid mechanics topics as an old science for applications of nanotechnology in science and medicine. The purpose of the book is introducing a new method for implementation and application of mechanical engineering governing equation in biological fluid and nanofluid flow and heat transfer. So introducing biomedical engineering and some applications as new branches of mechanical engineering (fluid mechanics and heat transfer applications) and nanotechnology is necessary as well as reviewing the history of fluid flow and heat transfer. Therefore, after reviewing the history of the book topics in this chapter, the thermodynamic principles and their application in human body and bio-system heat and mass transfer are discussed in Chapters 2 and 3, respectively. In Chapter 4 (fluid mechanics), biofluid flow (blood) is investigated. The properties and simulation of the bio-nano fluid is explained in Chapter 5. In the last chapter (Chapter 6), nanosystem application in drug delivery system as a new method in cancer therapy is discussed.

REFERENCES

[1] B.W. Carroll, Archimedes' Principle, Weber State University.
[2] A.Y. Hassan, Transfer of Islamic Technology to the West, Part II: Transmission of Islamic Engineering.
[3] J. Adamy, A. Flemming, Soft variable-structure controls: a survey, Automatica 40 (11) (November 2004) 1821–1844, http://dx.doi.org/10.1016/j.automatica.2004.05.017. Elsevier.
[4] D.R. Hill, Engineering, in: R. Rashed (Ed.), Encyclopedia of the History of Arabic Science, vol. 2, Routledge, London, New York, pp. 751–795 [776].
[5] M. Clagett, The Science of Mechanics in the Middle Ages, University of Wisconsin Press, 1961, p. 64.
[6] Encyclopædia Britannica, Al-Biruni (Persian Scholar and Scientist) — Britannica Online Encyclopedia. Britannica.com.
[7] K.C. Cheng, T. Fujii, Heat in history Isaac Newton and heat transfer, Heat Transfer Engineering 19 (4) (1998) 9–21.
[8] Professor Heinz Wolff. Heinzwolff.co.uk.
[9] MIT, Department of Biological Engineering.
[10] Oxford English Dictionary, third ed., November 2010.
[11] L. Waite, J. Fine, Applied Biofluid Mechanics, Mc-Graw Hill, 2007, http://dx.doi.org/10.1036/0071472177.
[12] K.E. Drexler, Engines of Creation: The Coming Era of Nanotechnology, Doubleday, 1986, ISBN 0-385-19973-2.
[13] K.E. Drexler, Nanosystems: Molecular Machinery, Manufacturing, and Computation, Wiley, 1992, ISBN 0-471-57518-6.
[14] N. Taniguchi, On the basic concept of 'nano-technology', in: Proceedings of the International Conference on Production Engineering, Tokyo, 1974, Part II, Japan Society of Precision Engineering, 1974.
[15] C. Toumey, Reading Feynman into nanotechnology: a text for a new science (PDF), Techné 13 (3) (2008) 133–168.
[16] G. Binnig, H. Rohrer, Scanning tunneling microscopy, IBM Journal of Research and Development 30 (4) (1986) 355–369.
[17] X. Qi Wang, S. Mujumdar, Heat transfer characteristics of nanofluids: a review, International Journal of Thermal Sciences 46 (2007) 1–19.
[18] M. Ghassemi, A. Shahidian, G. Ahmadi, S. Hamian, A new effective thermal conductivity model for a bio-nanofluid (blood with nanoparticle Al_2O_3), International Communications in Heat and Mass Transfer 37 (2010) 929–934.

CHAPTER 2

Thermodynamics

2.1 FUNDAMENTAL

Thermodynamics consists of two words: thermo (heat) and dynamics (power). It is a branch of science that deals with conversion of heat to work. It was established in the 19th century [1]. Historically, it dealt only with work generated by a hot body (heat engine) and efforts to make it a more efficient heat engine. Today, thermodynamics deals mostly with energy and its relationship between properties of substances.

Thermodynamics generally starts with several basic concepts and leads to different thermodynamics laws. A thermodynamic system is a quantity of matter, which is defined by its boundary. Everything outside the boundary is called the surroundings or environment. The environment often contains one or more idealized heat reservoirs—heat sources with infinite heat capacity enabling them to give up or absorb heat without changing their temperature. The boundary can be real or imaginary, fixed or movable. There are two types of systems: closed and open. A closed system (control mass) is a system with fixed quantity of matter. Thus, no mass crosses the boundary of the system. In an open system (control volume) the quantity of mass is not constant and mass can cross the boundary. An open system exchanges both matter and energy with its surroundings, whereas a closed system exchanges only energy with its surroundings. The isolated system exchanges neither energy nor matter with its surroundings. An example of a true isolated system is the universe with energy stored in it.

Each system is characterized by its properties. Thermodynamic properties are a macroscopic characteristic of a living entity to which a numerical value is assigned at a given time without knowledge of its *history*. *Properties* are either intensive (exist at a point in space, like temperature, pressure, and density) or extensive [depends on the size (or extent) of the system, like mass and volume]. There are a number of different intensive properties that are used to characterize material behavior. The three most important independent properties that usually describe a system are temperature, pressure, and specific volume [1]. Temperature is the measure of the relative warmth or coolness of a body. In other words, it is the intensity of heat in an object and is expressed mainly by a comparative scale and shown by a thermometer. Pressure is the amount of force that is exerted on a surface per unit area and specific volume is the number of cubic meters occupied by 1 kg of a particular substance.

The four laws of thermodynamics that describe the temperature (zeroth law of thermodynamics), energy (first law of thermodynamics), entropy (second law of

Nano and Bio Heat Transfer and Fluid Flow
ISBN 978-0-12-803779-9, http://dx.doi.org/10.1016/B978-0-12-803779-9.00002-9

thermodynamics), and entropy of substances at the absolute zero temperature (third law of thermodynamics) are discussed in detail in the following sections.

2.1.1 Zeroth Law of Thermodynamics

This law serves as a basis for the validity of temperature measurement. The Zeroth Law of Thermodynamics indicates that if two bodies are in thermal equilibrium with a third body, they are also in thermal equilibrium with each other. Replacing the third body with a thermometer helps measuring temperature of a system. Temperature is measured by means of a thermometer or other instruments having a scale calibrated in units called degrees. The size of a degree depends on the particular temperature scale being used.

2.1.2 The First Law of Thermodynamics

Based on experimental observation, energy can neither be created nor destroyed; it can only change forms. The first law of thermodynamics (or the conservation of energy principle) states that during an interaction between a system and its surroundings, the amount of energy gained by the system must be exactly equal to the amount of energy lost by the surroundings.

For a closed system (control mass), the first law of thermodynamics is shown as:

$$\begin{pmatrix} \text{Net amount of energy transfer as heat} \\ \text{and work to/or from the system} \end{pmatrix}$$

$$= \begin{pmatrix} \text{Net change in amount of enery} \\ \text{(increase/or decrease) with in the system} \end{pmatrix} \qquad (2.1)$$

$$\text{Or}$$

$$Q - W = \Delta E$$

In Eq. (2.1) Q and W are the net energy transfer by heat and work across the system boundary, respectively. The unit for Q and W is joule (J). ΔE is the net change of total energy within the system and its unit is joule (J). Total energy, E, of a system consists of internal energy (U), kinetic energy (KE), and potential energy (PE). The change in total energy of a system is expressed by:

$$\Delta E = \Delta U + \Delta KE + \Delta PE \qquad (2.2)$$

where:

$$\Delta U = (U_2 - U_1) \qquad (2.3)$$

Internal energy is represented by the symbol U, and $U_2 - U_1$ is the change in internal energy of a process.

$$\Delta KE = \frac{1}{2}m\left(V_2^2 - V_1^2\right) \tag{2.4}$$

where V is the magnitude of system velocity, $\frac{1}{2}mV^2$ is the kinetic energy, KE, of the body and ΔKE is the change in kinetic energy of the system.

$$\Delta PE = mg(z_2 - z_1) \tag{2.5}$$

Z is the magnitude of the system elevation relative to surface earth, mgz is the gravitational potential energy, PE, and the change in gravitational potential energy is ΔPE.

By inserting Eqs. (2.2)–(2.5) into Eq. (2.1), the first law of thermodynamics becomes:

$$Q - W = \Delta U + \frac{1}{2}m\left(V_2^2 - V_1^2\right) + mg(z_2 - z_1) \tag{2.6}$$

The instantaneous time rate form of the first law of thermodynamics is:

$$\frac{dE}{dt} = \dot{Q} - \dot{W} \tag{2.7}$$

where $\dot{Q}$ and $\dot{W}$ are the rates of heat and work transfer across the boundary, respectively.

The first law of thermodynamics for open system (control volume) is expressed as:

$$\left(\begin{array}{c}\text{Time rate of change of the energy} \\ \text{within the control volume}\end{array}\right) = \left(\begin{array}{c}\text{Net rate of energy transfer} \\ \text{as heat and work to/from} \\ \text{control volume at time } t\end{array}\right)$$
$$+ \left(\begin{array}{c}\text{Net rate of energy transfer} \\ \text{by mass entering the} \\ \text{control volume}\end{array}\right) \tag{2.8}$$

Or

$$\frac{dE_{cv}}{dt} = \dot{Q}_{cv} - \dot{W}_{cv} + \sum_i \dot{m}_i\left(h_i + \frac{V_i^2}{2} + gz_i\right) - \sum_e \dot{m}_e\left(h_e + \frac{V_e^2}{2} + gz_e\right)$$

where subscripts i and e denote the inlet to and exit from the system, respectively, h is the enthalpy, and $\dot{m}$ is the mass flow rate.

2.1.3 The Second Law of Thermodynamics

The second law of thermodynamics talks about the usefulness of energy as well as energy transfer direction. In other words, the second law of thermodynamics places a limit on the first law of thermodynamics. The second law states that the total system work is always less than the heat supplied into the system. For a closed system, the second law of thermodynamics is expressed as:

$$S_2 - S_1 = \int_1^2 \left(\frac{\delta Q}{T}\right)_b + S_{gen} \tag{2.9}$$

In Eq. (2.9) b is the system boundary, T is the absolute temperature, $\dot{Q}$ is the rate of energy transfer by heat, and S_{gen} is the amount of entropy generated by system irreversibility. By combining Eqs. (2.6) and (2.9), the irreversibility associated with a process (I) may be expressed by:

$$I = T_0 S_{gen} \tag{2.10}$$

The destruction of exergy due to irreversibility within the system is I and T_0 is the temperature of the surroundings. The second law of thermodynamics for a steady-flow system may be expressed as:

$$\frac{dS_{cv}}{dt} = \sum_j \frac{\dot{Q}_j}{T_j} + \sum_i \dot{m}_i s_i - \sum_e \dot{m}_e s_e + S_{gen} \tag{2.11}$$

Entropy in the microscopic or statistical view is a logarithm measure of the number of states (X_i) with significant probability of being occupied as given:

$$S = \sigma_B \ln X_i \tag{2.12}$$

where σ_B is the Boltzmann constant ($= 1.38 \times 10^{-23}$ m^2 kg/s^2 K) [2].

2.1.4 The Third Law of Thermodynamics

The third law of thermodynamics, formulated by Walter Nernst, also known as the Nernst heat theorem, is based on the studies of chemical reactions at low temperatures and specific heat measurements at temperatures approaching absolute zero. The third law of thermodynamics states that the entropy of substances is zero at the absolute zero of temperature. An example is pure crystalline substance that has zero entropy at the absolute zero of temperature, 0K.

2.1.5 Gibbs Free Energy

The Gibbs free energy (Gibbs energy or Gibbs function or free enthalpy to distinguish it from Helmholtz free energy) is a thermodynamic potential that measures the maximum or reversible work by a thermodynamic system at a constant temperature and pressure [3]. The Gibbs free energy (kJ), the maximum amount of nonexpansion work that can be adopted from a thermodynamically closed system, is the maximum that can be attained only in a completely reversible process.

The Gibbs free energy is defined as:

$$G(p, T) = U + pV - TS \quad \text{or} \quad \Delta G = \Delta H - T\Delta S \tag{2.13}$$

In Eq. (2.13), U, p, V, S, and H are the internal energy (J), pressure (pa), volume (m^3), temperature (K), entropy (J/K), and enthalpy (J), respectively.

The Gibbs free energy total differential natural variables may be derived via Legendre transforms of the internal energy.

$$dU = Tds - pdV + \sum_i \mu_i \, dN_i \tag{2.14}$$

Because S, V, and N_i are extensive variables, Euler's homogeneous function theorem allows easy integration of dU [4]:

$$U = TS - pV + \sum_i \mu_i \, N_i \tag{2.15}$$

2.2 THERMODYNAMICS OF SMALL SYSTEMS

2.2.1 Different Types of Thermodynamics

T.L. Hill developed the "Thermodynamics of small system" in 1960. He modified the theory of thermodynamic equilibrium of Boltzmann and Gibbs to develop nanothermodynamics. Nanothermodynamics, also known as thermodynamics of small systems, applies the standard thermodynamics to systems that are between 1 and 100 nm, systems that are about 1000 times smaller than the human cells. Nanothermodynamics extends the standard thermodynamics for small systems to facilitate finite-size effect on the scale of nanometers. This knowledge helps us to understand the overall thermodynamic equilibrium behavior of the small system.

Surface thermodynamics, nonextensive statistical mechanics, and Hill's approach based on the concept of subdivision potential are some of the approaches that are utilized to deal with the concept of nanothermodynamics.

As explained, thermodynamics describes the bulk behavior of the body, not the microscopic behaviors of the very large numbers of its microscopic constituents, such as molecules. The basic results of thermodynamics rely on the existence of idealized states of thermodynamic equilibrium.

The important branches of thermodynamics are Classical thermodynamics, Local equilibrium thermodynamics, Nonequilibrium thermodynamics, and Statistical thermodynamics. The plain term "thermodynamics" refers to a macroscopic description of bodies and processes [5]. Usually the plain term "thermodynamics" refers by default to equilibrium as opposed to nonequilibrium thermodynamics. The qualified term "statistical thermodynamics" refers to descriptions of bodies and processes in terms of the atomic or other microscopic constitution of matter, using statistical and probabilistic reasoning. Thermodynamic equilibrium is one of the most important concepts for thermodynamics [5,6]. In thermodynamic equilibrium there are no net macroscopic flows of matter or of energy, either within a system or between systems. The system is said to be in thermodynamic equilibrium if the Mechanical equilibrium, Chemical equilibrium, and Thermal equilibrium are satisfied.

2.2.1.1 Classical Thermodynamics

Classical thermodynamics accounts for the adventures of a thermodynamic system in terms either of its time–invariant equilibrium states or of its continually repeated cyclic processes, but, formally, not both in the same account. In classical thermodynamics, rates of change are not admitted as variables of interest. An equilibrium state stands endlessly without change over time, whereas a continually repeated cyclic process runs endlessly without a net change in the system over time.

2.2.1.2 Local Equilibrium Thermodynamics

Local equilibrium thermodynamics is concerned with the time courses and rates of progress of irreversible processes in systems that are smoothly spatially inhomogeneous. It admits time as a fundamental quantity, but only in a restricted way. Rather than considering time–invariant flows as long–term average rates of cyclic processes, local equilibrium thermodynamics considers time–varying flows in systems that are described by states of local thermodynamic equilibrium. Local equilibrium thermodynamics considers processes that involve the time–dependent production of entropy by dissipative processes, in which kinetic energy of bulk flow and chemical potential energy are converted into internal energy at time rates that are explicitly accounted for. Time–varying bulk flows and specific diffusional flows are considered, but they are required to be dependent variables, derived only from material properties described only by static macroscopic equilibrium states of small local regions. The independent state variables of a small local region are only those of classical thermodynamics.

2.2.1.3 Statistical Thermodynamics

Statistical thermodynamics, also called statistical mechanics, or statistical physics, emerged with the development of atomic and molecular theories in the second half of the 19th century and early 20th century. It provides an explanation of classical thermodynamics. It deals with average properties of the molecules, atoms, or elementary particles in random motion in a system of many such particles and relates these properties to the thermodynamic and other macroscopic properties of the system.

In 1878, Maxwell proposed the new term "statistical mechanics" [7].

Max Planck was the first who wrote down explicitly the famous formula as below [7]:

$$S = k \log W \qquad (2.16)$$

On the first glance the new field of physics developed by Boltzmann and Gibbs, called "statistical mechanics," "statistical thermodynamics," or "statistical physics," was beautiful, attractive to young scientists, and very perspective [7].

2.2.1.4 Nonequilibrium Thermodynamics

Nonequilibrium thermodynamics is a branch of thermodynamics that deals with physical systems that are not in thermodynamic equilibrium but can be adequately described in terms of variables (nonequilibrium state variables) that represent an extrapolation of the variables used to specify the system in thermodynamic equilibrium. Nonequilibrium thermodynamics is concerned with transport processes and with the rates of chemical reactions. It relies on what may be thought of as more or less nearness to thermodynamic equilibrium. Nonequilibrium thermodynamics is a work in progress, not an established edifice. This article will try to sketch some approaches to it and some concepts important for it.

Almost all systems found in nature are not in thermodynamic equilibrium, for they are changing or can be triggered to change over time, and are continuously and discontinuously subject to flux of matter and energy to and from other systems and to chemical reactions. Some systems and processes are, however, in a useful sense, near enough to thermodynamic equilibrium to allow description with useful accuracy by currently known nonequilibrium thermodynamics. Nevertheless, many natural systems and processes will always remain far beyond the scope of nonequilibrium thermodynamic methods. This is because of the very small size of atoms, as compared with macroscopic systems.

Nonequilibrium thermodynamics has been successfully applied to describe biological processes such as protein folding/unfolding and transport through membranes [7]. Also, ideas from nonequilibrium thermodynamics and the informatic theory of entropy have been adapted to describe general economic systems [8,9].

2.2.2 Nanothermodynamics

Nanoscale system shows a wide range of special physical and chemical properties not observed in macroscopic materials. One of the characteristic features of nanosystems is their high surface-to-volume ratio (As/V). The knowledge of As/V is necessary to explain the thermodynamic principles governing nanoscale systems. Due to the large value of As/V its properties depend on the size, which is not the case for ordinary large systems. As particle size decreases the As/V increases so much that the change in melting temperature becomes readily noticeable. Low melting temperature is one of the unique properties nanosystems. For systems greater than 100 nm the As/V and temperature change remains relatively close to zero. Melting point and As/V of systems below 1 nm have no physical meaning as these sizes are governed by subatomic and single atom phenomenon.

In general, very small systems have much more of their matter exposed at the surface, which has severe implications concerning the particle's surface tension and atomic bonding. The majority of atoms in nanosystems are surface atoms, which means that nanosystems have a much greater surface energy than bulk materials. Surface tension, the rate at which work must be done on a surface to expand its area, is an important parameter that changes the melting temperature of a nanostructure. The change in melting temperature of a substance (ΔT) is determined by [10]:

$$\Delta T = -2\frac{\sigma \overline{V} T_m}{r \Delta \overline{H}_m} \tag{2.17}$$

where σ is the particle surface tension, $\overline{V}$ is the molar volume, T_m is the normal melting temperature of bulk material, r is the particle radius, and $\Delta \overline{H}_m$ is the change in molar enthalpy and is given by:

$$\Delta \overline{H}_m = T_m \Delta \overline{S}_m \tag{2.18}$$

where $\Delta \overline{S}_m$ is the molar entropy of fusion, the rate at which energy would have to be added to a substance to raise its temperature per mole of substance at the material's melting point, and is determined by:

$$\Delta \overline{S}_m = -\frac{\mu_m}{\Delta T} \tag{2.19}$$

where μ_m is the chemical potential for melting the particle. The chemical potential of a nanoparticle is influenced by surface tension, whereas the surface tension and surface energy depend on the nanosystem size. The chemical potential (μ) is a form of potential energy that can be absorbed or released during chemical reaction and is given by:

$$\mu = \frac{\Delta G}{N} \tag{2.20}$$

where ΔG is Gibbs free energy and N is number of moles of the substance. For pure substance the chemical potential is equivalent to the Gibbs function, $G = N\mu$. For instance, the chemical potential of a sphere is:

$$\mu_s = \frac{\Delta G}{N} = \frac{2\overline{V}\sigma}{r} \tag{2.21}$$

where $N = V/\overline{V}$ and $rdA = 2dV$.

2.2.3 Thermodynamic Properties of Small Systems

The partition function is an important parameter that is the link between the thermodynamic coordinates of a macroscopic system and the coordinates of the constituent microscopic systems. The total partition function (Z) is defined as below [11]:

$$Z = Z^N = \left[\sum \exp(-\beta \varepsilon_i) \right]^N \tag{2.22}$$

since $\beta = 1/kT$ and ε is the particle energy.

The internal energy of system comprising N atoms or molecules is equal to the sum of the energies of the individual particles. This parameter is obtained as follows [11]:

$$U = \sum N_i \varepsilon_i = N\overline{\varepsilon_i} \tag{2.23}$$

which $\overline{\varepsilon}$ is calculated by:

$$\overline{\varepsilon} = -\frac{1}{Z}\left(\frac{\partial Z}{\partial \beta}\right)_V = -\left(\frac{\partial \ln Z}{\partial \beta}\right)_V \tag{2.24}$$

The entropy as a function of the total partition function is calculated as [11]:

$$U = \sum N_i \varepsilon_i = N\overline{\varepsilon_i} \tag{2.25}$$

In addition, pressure (P), Gibbs function (G), and enthalpy (H) are defined by bellow equations [11]:

$$P = n\,RT\left(\frac{\partial \ln Z}{\partial V}\right)_T \tag{2.26}$$

$$G = -n\,RT\left[\ln\left(\frac{Z}{N}\right) + 1 - V\left(\frac{\partial \ln Z}{\partial V}\right)_T\right] \tag{2.27}$$

$$H = n\,RT\left[\left(\frac{\partial \ln Z}{\partial \ln T}\right)_V + \left(\frac{\partial \ln Z}{\partial \ln V}\right)_T\right] \tag{2.28}$$

2.3 THERMODYNAMICS OF BIOSYSTEMS

2.3.1 Biological Thermodynamics

Biological thermodynamics is the quantitative study of the energy transductions that occur in and between living organisms, structures, and cells and function of the chemical processes underlying these transductions. Biological thermodynamics may address the question of whether the benefit associated with any particular phenotypic trait is worth the energy investment it requires.

The book *Energy Transformations in Living Matter* was the first major publication on the thermodynamics of biochemical reactions by Hans Kornberg in 1957 [12]. In addition, the appendix contained the first-ever published thermodynamic tables, written by Kenneth Burton, to contain equilibrium constants and Gibbs free energy of formations for chemical species, to be able to calculate biochemical reactions [12].

The field of biological thermodynamics is focused on principles of chemical thermodynamics in biology and biochemistry. Principles covered include the first law of thermodynamics, the second law of thermodynamics, Gibbs free energy, statistical thermodynamics, reaction kinetics, and hypotheses of the origin of life. Presently, biological thermodynamics concerns itself with the study of internal biochemical dynamics such as: ATP hydrolysis, protein stability, DNA binding, membrane diffusion, enzyme kinetics [13], and other such essential energy-controlled pathways. In terms of thermodynamics, the amount of energy capable of doing work during a chemical reaction is measured quantitatively by the change in the Gibbs free energy. The physical biologist Alfred Lotka [14] attempted to unify the change in the Gibbs free energy with evolutionary theory.

Biological systems, which live in symbiosis with their environment, are formed and maintained by the dissipative processes that exchange energy between the system and its environment. Energy transformations in biology are dependent primarily on photosynthesis.

Some living organisms like plants need sunlight directly, whereas other organisms like humans can acquire energy from the sun indirectly [15]. The relationship between the energy of the incoming sunlight and its wavelength λ or frequency ν is given by:

$$E = hc/\lambda = h\nu \tag{2.29}$$

where h is the Planck constant (6.63×10^{-34} J s) and c is the speed of light (2.998×10^8 m/s). Plants trap this energy from the sunlight and undergo photosynthesis, effectively converting solar energy into chemical energy. To transfer the energy once again, animals will feed on plants and use the energy of digested plant materials to create biological macromolecules.

In living organisms, the main repositories of energy are macromolecules, which store energy in the form of covalent and noncovalent chemical bonds, and unequal concentrations of solutes, principally ions, on opposite sides of a cell membrane. The other types of energy distribution are shown in Table 2.1 [15].

All bodily functions, from thinking to lifting weights, require energy. Our own bodies, like all living organisms, are energy conversion machines. Conservation of energy implies that the chemical energy stored in food is converted into work, thermal energy, or stored as chemical energy in fatty tissue. By far the largest fraction goes to thermal

Table 2.1 Different types of kinetic and potential energy distribution in cells [4]

Kinetic energy	Potential energy
Heat or thermal energy: Energy of molecular motion in all organisms. At 25°C this is about 0.5 kcal/mol	*Bond energy:* Energy of covalent and noncovalent bonds, for example, a σ bond between two carbon atoms or van der Waals interactions. These interactions range in energy from as much as 14 kcal/mol for ion—ion interactions to as little as 0.01 kcal/mol for dispersion interactions; they can also be negative, as in the case of ion—dipole interactions and dipole—dipole interactions
Radiant energy: Energy of photons, for example, in photosynthesis. The energy of such photons is about 40 kJ/mol	*Chemical energy:* Energy of a difference in concentration of a substance across a permeable barrier, for instance, the lipid bilayer membrane surrounding a cell. The magnitude depends on the difference in concentration across the membrane. The greater the difference, the greater the energy
Electrical energy: Energy of moving charged particles, for instance, electrons in reactions involving electron transfer. The magnitude depends on how quickly the charged particle is moving. The higher the speed, the greater the energy	*Electrical energy:* Energy of charge separation, for example, the electrical field across the two lipid bilayer membranes surrounding a mitochondrion. The electrical work required to transfer monovalent ions from one side of a membrane to the other is about 20 kJ/mol

energy, although the fraction varies depending on the type of physical activity. The fraction going into each form depends both on how much we eat and on our level of physical activity. Fig. 2.1 shows the first law of thermodynamics in body.

The kidneys and liver consume a surprising amount of energy, but the biggest surprise of all is that a full 25% of all energy consumed by the body is used to maintain electrical

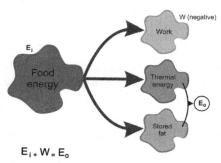

$$E_i + W = E_o$$

Figure 2.1 The first law of thermodynamics in body.

potentials in all living cells. (Nerve cells use this electrical potential in nerve impulses.) This bioelectrical energy ultimately becomes mostly thermal energy, but some is utilized to power chemical processes such as in the kidneys and liver, and in fat production. The rate at which the body uses food energy to sustain life and to do different activities is called the metabolic rate, and the corresponding rate when at rest in a neutrally temperate environment, in the postabsorptive state, is called the basal metabolic rate (BMR). About 75% of food calories are used to sustain basic body functions included in the BMR. The energy included in BMR is divided among various systems in the body, with the largest fraction going to the liver, spleen, and brain. The BMR is a function of age, gender, total body weight, and amount of muscle mass (which burns more calories than body fat). Work done by a person (useful work) is work done on the outside world, such as lifting weights. This is accomplished by exerting forces on the outside world. Forces exerted by the body are nonconservative, so that they can change the mechanical energy ($KE + PE$) of the system worked upon, and this is often the goal. But the work done by the heart when pumping blood is internal work.

2.3.2 Thermodynamics of Living Cells

All living organisms on earth are made up of cells. Cells, the basic structural unit for all organisms, are the small compartments that hold the biological equipment necessary to keep an organism alive and successful. Living cells stay alive by using and transforming energy that are in forms of thermal, mechanical, chemical, electrical, electromagnetic, and atomic energy. Transformation of energy involves chemical reactions, which is based on thermodynamics.

Metabolism is the chemical life of a cell. Metabolism is the sum total of all the chemical reactions, also called pathways, taking place in an organism. The chemical reactions are either catabolic (from larger to smaller molecules or low oxidative state to high oxidative state) or anabolic (from smaller to larger molecules or high oxidative state to low oxidative state). All catabolic reactions are oxidative and produce energy, whereas the biosynthetic reactions are reductive and consume energy. Biosynthetic reactions are essential for the cellular material formation.

A living system is not an isolated system. It gets the nutrients from the surrounding and exchanges heat or matter with the environment. Therefore cell behaves as an open system that is not in equilibrium. The cell as an open system with its surrounding causes increases in entropy or disorder.

In considering thermodynamic relationships in biological reactions the main interest is in reactions that approach equilibrium and have maximum entropy. The maximum entropy, the driving force of all processes, is reached by giving up heat to absorbing heat from the surrounding. The relation between heat and entropy is given by Eq. (2.13).

Cell metabolism consists of sequences of oxidation and reduction reactions. Oxidation is generally defined as the loss of electrons and reduction as the gain of electrons. Electrons released by oxidation are accepted by an oxidizing agent and establish an electric current. The tendency of substances to donate or accept electrons leads to the calculation of Gibbs free energy changes for oxidation–reduction reactions. This is called oxidation–reduction potential. The Gibbs free energy change associated with an oxidative reaction is calculated from the Nernst equation as follows [16]:

$$-\Delta G = NF\Delta \epsilon \tag{2.30}$$

where N represents the number of electrons involved in the reaction, F is the Faraday constant (=96,494 coulombs) necessary to convert one equivalent of ions, and the potential difference, $\Delta \epsilon$, is expressed by the Nernst equation as [16]:

$$\Delta \epsilon = \epsilon_h - \epsilon_0 = \left(\frac{RT}{nF}\right)\ln K \tag{2.31}$$

where ϵ_h is the measured oxidation–reduction potential difference, ϵ_0 is the standard electrode potential, and K is the equilibrium constant (electromotive force) of a reversible reaction. The electromotive force is the algebraic difference of the two half-cell potentials.

Energy coupling, the use of an exergonic process to drive an endergonic process, is very common in biological systems. The exergonic reactions are spontaneous and lead to a decrease in Gibbs free energy (negative value of ΔG) and an increase in entropy (positive value of ΔS). The endergonic reactions are not spontaneous and lead to an increase in Gibbs free energy (positive value of ΔG) and a decrease in entropy (positive value of ΔS). Hydrolysis of adenosine triphosphate compound, ATP, is the most common coupling reaction in which ATP is hydrolyzed to ADP and inorganic phosphate (Pi). In cells, the ATP hydrolysis reaction takes place by a process known as cellular energy metabolism, which is far away from chemical equilibrium. Energy metabolism is the catabolism of organic molecules that yield ATP and other useful forms of chemical energy.

2.4 ENERGY AND EXERGY ANALYSIS OF HUMAN BODY

Human body core temperature is maintained a constant, which is known as homeothermy. The temperature of the body peripheral parts, such as hands and feet, may vary as the surrounding temperature varies. For this reason the human body thermal behavior is divided into: the inner section (also known as the body core) and the outer section (also known as the skin layer); see Fig. 2.2.

The body core and skin compartments exchange energy passively through direct contact and dynamically through the thermoregulatory controlled compartments. The

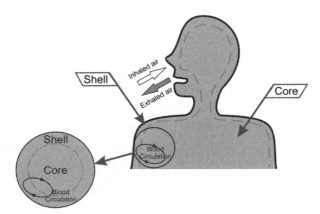

Figure 2.2 Modeling of human body consisting of two subsystems.

energy and mass for the human body's vital processes are exchanged with the environment. Thermal comfort is a state of mind that describes contentment with the environment and is adjusted by the person. The body exchanges heat with the environment of the room due to the human body's metabolism and activity.

The metabolic rate generally slows as a person ages. This is partly because of the loss of muscle tissue, and hormonal and neurological changes. For example, metabolism in babies and children speeds up as they go through periods of growth. The metabolism as a function of age is indicated in Fig. 2.3.

As known the human body acts as an open system (i.e., a heat engine). The energy conversion and exergy analysis of such system is obtained by applying the first and second laws of thermodynamics to an open system. Depending on the human body parameters (such as age, metabolism, muscle mass, body size, gender, physical activity, hormonal factors, drugs, and diet) and environmental factors, the second law determines the exergy

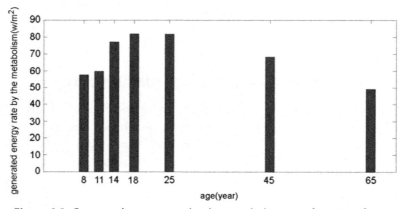

Figure 2.3 Generated energy rate by the metabolism as a function of age.

consumption within the human body. Energy generated by metabolism is an important factor affecting entropy generation in the body. Human metabolism itself is affected by human age.

The exergy analysis considers the quantity and quality of the energy that is supplied to the human body. All human organisms are generally influenced by the climatic factors and usually adjust themselves to the environmental conditions. Human life is a function of entropy and a biological organism tends to reach the minimum entropy production [17]. It is shown that the exergy consumption by human body exhibits a minimal value at certain combinations of environmental parameters [18]. The lowest human body exergy consumption is associated with thermal sensation votes close to neutrality, tending to a slightly cool sensation [19]. The human body operates more efficiently at low humidity and high temperatures [20]. The following section presents the equations that determine the energy and exergy consumption for human body.

2.4.1 Energy Analysis

The human body transfers heat by conduction, convection, and radiation to the environment. It also produces heat by metabolism and loses heat generally by evaporation and diffusion of body liquids. Fig. 2.4 shows the schematic of the heat transfer of the human body with indoor and outdoor environments.

The general energy balance equation of the human body is:

$$En_M = En_W + En_{dif} + En_{sw} + En_{res} + En_{loss,\Delta T} + \sum En_{Hx} + En_s \qquad (2.32)$$

where E is energy and its unit is W/m^2.

The effect of age on metabolism rate of energy generation, En_M, is calculated by:

$$En_M = En_{shiv} + En_{act} \qquad (2.33)$$

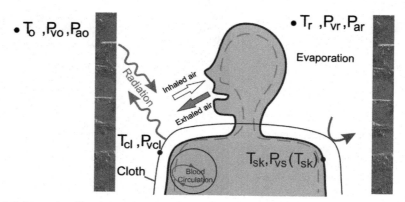

Figure 2.4 Schematic of heat transfer of the human body with indoor and outdoor environments.

where energy of shivering, En_{shiv}, and energy of activity, En_{act}, are given by:

$$\dot{En}_{shiv} = 19.4 \times (34 - T_{sk})(37 - T_{cr})$$
$$\dot{En}_{act} = 58 \tag{2.34}$$

The external work rate is zero, $\dot{En}_w = 0$. The energy loss rate by water vapor diffusion through the skin, En_{dif}, the energy loss rate by evaporation of sweat, En_{sw}, and the energy loss rate by respiration, En_{res}, are given by, respectively:

$$En_{dif} = (0.0035)\left(P_{sv,T_{sk}} - P_{v,a}\right)$$
$$En_{res} = (0.0000172)En_M(5867 - P_{va})$$
$$En_{sw} = (0.42)(En_M - En_w - 58.15) \tag{2.35}$$

The energy loss rate due to difference in temperature, $En_{loss,\Delta T}$, and the total heat exchange rate, En_{hx}, are determined by:

$$En_{loss,\Delta T} = (0.0014)En_M(T_{exh} - T_{inh})$$
$$\sum En_{Hx} = \frac{(T_{sk} - T_{cl})}{(0.155)I_{cl}} \tag{2.36}$$

Moreover, En_s is the stored exergy rate in the body.

The energy loss rate distribution for three ages (11, 18, and 45 years) is shown in Fig. 2.5.

In addition, the general thermal sensation is obtained using predicted mean vote (PMV) index based on ISO7730 standards [21]. The PMV index predicts the main value of the people on the seven-point thermal sensation scale as $+3$, $+2$, $+1$, 0, -1, -2, -3

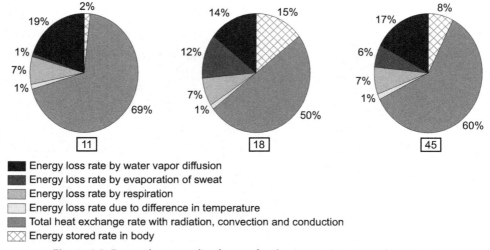

Figure 2.5 Energy loss rate distribution for three ages (11, 18, and 45 years).

with the meaning of hot, warm, slightly warm, neutral, slightly cool, cool, cold, respectively.

$$PMV = (3.155)\left(0.303e^{-0.114(\dot{E}n_M)} + 0.028\right)L_{PMV} \tag{2.37}$$

$$L_{PMV} = \left(\dot{E}n_M \pm \dot{E}n_w\right) - \dot{E}n_{Diff} - \dot{E}n_{Sw} - \dot{E}n_{Res} - \dot{E}n_{loss,\Delta T} - \sum \dot{E}n_{Hx} \tag{2.38}$$

$$PPD = 100 - 95\left[e^{-0.03353(PMV)^4 - 0.2179(PMV)^2}\right]$$

The predicted mean vote as a function of age is shown in Fig. 2.6.

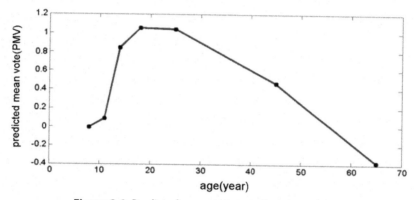

Figure 2.6 Predicted mean vote as a function of age.

2.4.2 Exergy Analysis

Six parameters such as metabolic thermal energy emission rate, clothing insulation level of human body, surrounding relative humidity, air velocity, air temperature, and mean radiant temperature are used to model the human body exergy consumption rates.

To determine the exergy consumption of the human body, first the core and the skin and the clothing surface temperatures are determined by a two node. This is done by the two personal parameters and the four objective parameters. Then the input—output exergy fluxes and exergy consumption rate of the human body are calculated using the values that are obtained for core, skin, and clothing temperatures.

The average weight and length used for analysis of the human rate of exergy consumption is given in Table 2.2.

In addition, the human body surface is determined by [22]:

$$A = \sqrt{\frac{l \times m}{36}} \tag{2.39}$$

Other related data are listed in Table 2.3.

Table 2.2 Human data used for analysis

	Age (*a*)	Weight (*m*)	Length (*l*)	Body surface (*A*)	Activity level (*act*)
1	8	28	132	1.0132	1
2	11	37	147	1.2292	1
3	14	50	162	1.5000	1.2
4	18	65	162	1.7927	1.2
5	25	71	178	1.8736	1.2
6	45	70	178	1.8604	1.2
7	65	74	178	1.9128	1

The general system exergy balance is given by:

$$[\text{Exergy consumed}] = [\text{Exergy input}] - [\text{Exergy stored}] + [\text{Exergy output}]$$

The human body thermal exergy balance consists of the core and the shell and is determined by:

$$Ex_M + Ex_{gen,cr} + Ex_{gen,mb-da} + Ex_{inh,air} + Ex_{abs,sk-cl}$$
$$= Ex_{rad,dc} + Ex_{conv} + Ex_{exh,air} + Ex_{sw,ha} + Ex_{stored,cr} + Ex_{stored,sk} + Ex_{cons} \quad (2.40)$$

The core liquid water exergy rate generated by metabolism, $EX_{gen,cr}$, and the stored exergy rate, $Ex_{stored,cr}$, are given by [22]:

$$Ex_{gen,cr} = V_{w,core}\rho_w \left\{ c_{p,w} \left[(T_{cr} - T_0) - T_0 \left(\ln \frac{T_{cr}}{T_{cr}} \right) \right] + \frac{R}{X_w} T_0 \left(\ln \frac{P_{sv,T_0}}{P_{vo}} \right) \right\} \quad (2.41)$$

$$Ex_{stored,cr} = \dot{Q}_{core} \left(1 - \frac{T_0}{T_{cr}} \right) \quad (2.42)$$

Table 2.3 System description data

T_a	24°C
RH_o	%30
T_{ra}	26°C
RH_r	%40
V_{air}	1 m/s
T_m	28°C
T_{cl}	28.86°C
T_{cr}	36.82°C
T_{sk}	33.96°C
Clo	0.8

The shell liquid water/dry air exergy rate generated by metabolism, $Ex_{gen,mb-da}$, the stored exergy rate, $Ex_{stored,sk}$, and the discharged radiant exergy rate, $Ex_{rad,dc}$, are determined by [22]:

$$Ex_{gen,mb-da} = V_{w,shell}\rho w \left\{ c_{p,w} \left[(T_{sk} - T_0) - T_0 \left(\ln \frac{T_{sk}}{T_{cr}} \right) \right] \right.$$
$$\left. + \frac{R}{X_w} T_0 \left[\left(\ln \frac{P_{sv,T_0}}{P_{vo}} \right) + \frac{P - P_{vr}}{P_{vr}} \left(\ln \frac{P - P_{vr}}{P - P_{v0}} \right) \right] \right\} \quad (2.43)$$

$$Ex_{stored,sk} = \dot{Q}_{sk} \left(1 - \frac{T_0}{T_{sk}} \right) \quad (2.44)$$

$$Ex_{stored,sk} = f_{ef}f_{cl}a_i\varepsilon_{cl}h_{rb} \frac{(T_{cl} - T_0)^2}{(T_{cl} + T_0)} \quad (2.45)$$

The metabolism exergy rate, Ex_m, is given by [22]:

$$Ex_M = En_M \left(1 - \frac{T_0}{T_{cr}} \right) \quad (2.46)$$

The inhaled and exhaled humid air exergy rates are given by, respectively [22]:

$$Ex_{inh,air} = V_{in} \left\{ \left[c_{p,a} \left(\frac{X_{da}}{RT_{ra}} \right) (P - P_{vr}) + c_{pa} \left(\frac{X_w}{RT_{ra}} \right) P_{vr} \right] \times \left[(T_{ra} - T_0) \right. \right.$$
$$\left. - T_0 \left(\ln \frac{T_{ra}}{T_0} \right) \right] + \frac{T_0}{T_{ra}} (P - P_{vr}) \left(\ln \frac{P - P_{ra}}{P - P_{v0}} \right) + P_{vr} \left(\ln \frac{P_{vr}}{P_{vo}} \right) \right\}$$
$$(2.47)$$

$$Ex_{inh,air} = V_{out} \left\{ \left[c_{p,a} \left(\frac{X_{da}}{RT_{cr}} \right) (P - P_{vr,T_{cr}}) + c_{pv} \left(\frac{X_w}{RT_{cr}} \right) P_{vr,T_{cr}} \right] \times \left[(T_{cr} - T_0) \right. \right.$$
$$\left. - T_0 \left(\ln \frac{T_{cr}}{T_0} \right) \right] + \frac{T_0}{T_{cr}} (P - P_{vr,T_{cr}}) \left(\ln \frac{P - P_{cr,T_{cr}}}{P - P_{v0}} \right)$$
$$+ P_{vr,T_{cr}} \left(\ln \frac{P_{vr,T_{cr}}}{P_{vo}} \right) \right\}$$
$$(2.48)$$

Furthermore, the body surface discharged radiant exergy rate, $Ex_{rad,dc}$, the convection exergy rate to the air, Ex_{conv}, and the water vapor/air exergy rate from sweat, $Ex_{sw,ha}$, are determined by [22]:

$$Ex_{rad,dc} = f_{ef}f_{cl}a_i\varepsilon_{cl}h_{rb} \frac{(T_{cl} - T_0)^2}{(T_{cl} + T_0)} \quad (2.49)$$

$$Ex_{conv} = f_{cl}f_{ccl}(T_{cl} - T_{ra}) \left(1 - \frac{T_0}{T_{cl}} \right) \quad (2.50)$$

Table 2.4 Exergy output analysis for different ages of human body

Exergy age (years)	$\dot{Ex}_{rad,dc}$	$\dot{Ex}_{conv}$	$\dot{Ex}_{exh,air}$	$\dot{Ex}_{sw,ha}$	$\dot{Ex}_{stored}$	$\dot{Ex}_{cons}$
8	0.006	0.1113	0.1962	0.0127	−0.0011	2.4620
11	0.006	0.1113	0.1962	0.0127	0.0201	2.5016
14	0.006	0.1113	0.1962	0.0127	0.1932	2.8241
18	0.006	0.1113	0.1962	0.0127	0.2411	2.9133
25	0.006	0.1113	0.1962	0.0127	0.2382	2.9079
45	0.006	0.1113	0.1962	0.0127	0.1056	2.6609
65	0.006	0.1113	0.1962	0.0127	−0.0836	2.3084

$$Ex_{sw,ha} = V_{w,shell}\rho w \left\{ c_{p,v}\left[(T_{cl} - T_0) - T_0 \ln\left(\frac{T_{cl}}{T_0}\right)\right] + T_0\left(\frac{R}{X_w}\right)\left[\ln\left(\frac{P_{vr}}{P_{vo}}\right)\right.\right.$$
$$\left.\left. + \frac{P - P_{vr}}{P_{vr}}\left(\ln\frac{P - P_{vr}}{P - P_{v0}}\right)\right]\right\} \tag{2.51}$$

Also "$\dot{Ex}_{cons}$" is the exergy consumption rate.

$$\dot{Ex}_{abs,sk-cl} = f_{ef}f_{cl}a_i\varepsilon_{cl}h_{rb}\frac{(T_i - T_0)^2}{(T_i + T_0)} \tag{2.52}$$

Exergy output rates of the human body are shown in Table 2.4.

As shown in Table 2.4 the discharged radiant exergy rate, the convection exergy rate, the exhaled humid air exergy rate, and the water vapor/air exergy rate from sweat are the same for all ages. However, the exergy stored and consumption are different and are high during the youth stages.

Fig. 2.7 shows the exergy output distribution for the human ages 11, 18, and 45 years.

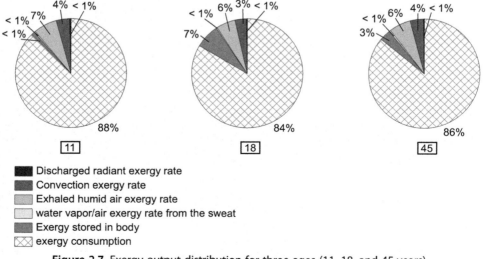

Figure 2.7 Exergy output distribution for three ages (11, 18, and 45 years).

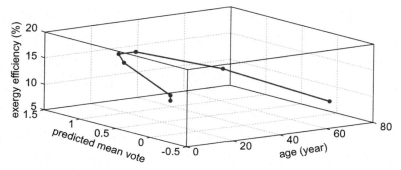

Figure 2.8 Exergy efficiency as a function of age and predicted mean vote.

Table 2.5 Exergy efficiency and predicted mean vote (PMV) analysis results of human body

Age (years)	PMV	Exergy efficiency
8	−0.0050	11.6661
11	0.0887	12.1626
14	0.8443	15.5366
18	1.0528	16.3004
25	1.0402	16.2558
45	0.4628	13.9644
65	−0.3791	9.5136

The exergy stored in the body is higher during the youth stage because higher energy gets stored in the body. The negative exergy stored results in negative predicted mean. Fig. 2.8 shows the exergy efficiency as a function of age and PMV.

Table 2.5 shows the exergy efficiency as a function of age and PMV. As expected the exergy efficiency is high for young people and it is lower for children and older people. The minimum and maximum exergy efficiencies are obtained for ages 65 and 18 years, respectively. Clothing and insulation increases the exergy efficiency.

REFERENCES

[1] Y.A. Cengel, M.A. Boles, Thermodynamics, an Engineering Approach, fifth ed.
[2] K. Stowe, An Introduction to Thermodynamics and Statistical Mechanics, second ed., Cambridge University Press, 2007.
[3] W. Greiner, L. Neise, H. Stöcker, Thermodynamics and Statistical Mechanics, Springer-Verlag, 1995, p. 101.
[4] W.R. Salzman, "Open Systems". Chemical Thermodynamics, University of Arizona, August 21, 2001. Archived from the original on 2007-07-07.
[5] B.C. Eu, Generalized Thermodynamics. The Thermodynamics of Irreversible Processes and Generalized Hydrodynamics, Kluwer Academic Publishers, Dordrecht, 2002, ISBN 1-4020-0788-4.

[6] R. Wildt, Thermodynamics of the gray atmosphere. IV. Entropy transfer production, Astrophysical Journal 174 (1) (1972) 69—77.

[7] W. Ebeling, I.M. Sokolov, Statistical thermodynamics and stochastic theory of nonequilibrium systems, Series on Advances in Statistical Mechanics, vol. 8, World Scientific.

[8] V. Pokrovskii, Econodynamics, The Theory of Social Production, Springer, Dordrecht, Heidelberg, London, New York, 2011. http://www.springer.com/physics/complexity/book/978-94-007-2095-4.

[9] J. Chen, The Unity of Science and Economics: A New Foundation of Economic Theory, Springer, 2015. http://www.springer.com/us/book/9781493934645.

[10] Z. Hua Li, D.G. Truhlar, Nanothermodynamics of metal nanoparticles, Chemical Science 5 (2014) 2605—2624.

[11] J.F. Lee, F.W. Sears, D.L. Turcotte, Statistical Thermodynamics, second ed., Addison-Wesley Publishing Company, Inc.

[12] R. Alberty, A short history of the thermodynamics of enzyme-catalyzed reactions, Journal of Biological Chemistry 279 (27) (2004) 27831—27836, http://dx.doi.org/10.1074/jbc.X400003200.

[13] A. Ito, T. Oikawa, Global mapping of terrestrial primary productivity and light-use efficiency with a process-based model, in: M. Shiyomi et al. (Eds.), Global Environmental Change in the Ocean and on Land (PDF), pp. 343—358.

[14] M.J. Farabee, Reactions and Enzymes. On-line Biology Book, Estrella Mountain Community College.

[15] D.T. Haynie, Biological Thermodynamics, Cambridge University Press, 2001, pp. 1—16.

[16] H.W. Doelle, Cell thermodynamics and energy metabolism, Biotechnology — Vol. I — Cell Thermodynamics and Energy Metabolism, Encyclopedia of Life Support Systems (EOLSS).

[17] I.T. Prigogine, J. Wiam, Biologic thermodynamique des phenomenes irreversibles, Experimenta 2 (1946) 451—453.

[18] M. Prek, V. Butala, Principles of exergy analysis of human heat and mass exchange with the indoor environmental, International Journal of Mass Transfer 48 (2010) 731—739.

[19] A. Simone, B.W. Olesen, An experimental study of thermal comfort at different combinations of air and mean radiant temperature, in: Healthy Buildings Conferences (HB 2009), 2009.

[20] C.E.K. Mady, S. Olivera Jr., Human body exergy metabolism, in: International Conference on Efficiency, Cost, Optimization, Simulation, and Environmental Impact of Energy Systems, Perugia, Italy, 2012, pp. 160-1—160-13.

[21] ISO7730, Ergonomics of the Thermal Environment Analytical Determination and Interpretation of Thermal Comfort Using Calculation of the PMV and PPD Indices and Local Thermal Comfort Criteria, International Standard Organization for Standardization, Geneva, 2005.

[22] A. Shaihidian, Z. Abbasi, Investigation of effective parameters on the human body exergy and energy model, in: 7th International Exergy, Energy and Environment Symposium, France, 2015.

CHAPTER 3

Biosystems Heat and Mass Transfer

3.1 INTRODUCTION

Heat transfer deals with the rate of heat transfer between physical systems. Heat transfer processes set limits to the performance of environmental components and systems. The transfer of heat is always from the higher temperature medium to the lower temperature medium. Therefore a temperature difference is required for heat transfer to take place. The rate of heat transfer also depends on the properties of the intervening medium through which the heat is transferred. Heat transfer through living biological tissue involves heat conduction in solid tissue matrix and blood vessels, blood perfusion (convective heat transfer between tissue and blood), cooling of human body by radiation, as well as metabolic heat generation. Heat transfer processes are classified into three types: conduction, convection, and radiation.

Conduction heat transfer is the transfer of heat through matter (i.e., solids, liquids, or gases) without bulk motion of the matter. In other words, conduction is the transfer of energy from the more energetic to less energetic particles of a substance due to interaction between the particles. For example, heat conduction can occur through wall of a vein in human body. The inside surface, which is exposed to blood, is at a higher temperature than the outside surface.

Convection heat transfer is due to the moving fluid. The fluid can be a gas or a liquid; both have applications in bio and nano heat transfer. In convection heat transfer, the heat is moved through bulk transfer of a fluid with nonuniform temperature. An example of such heat transfer is the flow of blood inside the human body and different organs.

Radiation heat transfer is the energy that is emitted by matter in the form of photons or electromagnetic waves. Radiation can take place through space without the presence of matter. In fact radiation heat transfer is highest in vacuum environment. Radiation can be important even in situations in which there is an intervening medium. An example is the heat transfer that takes place between a living entity with its surrounding.

3.1.1 Conduction Heat Transfer

Conduction heat transfer is the transfer of heat by means of molecular excitement within a material without bulk motion of the matter. Conduction heat transfer in gases and liquids is due to the collisions and diffusion of the molecules during their random motion. On the other hand, heat transfer in solids is due to the combination of lattice vibrations of the molecules and the energy transport by free electrons.

Nano and Bio Heat Transfer and Fluid Flow
ISBN 978-0-12-803779-9, http://dx.doi.org/10.1016/B978-0-12-803779-9.00003-0

To examine conduction heat transfer, it is necessary to relate the heat transfer to mechanical, thermal, or geometrical properties. Consider steady-state heat transfer through the wall of an aorta with thickness Δx where the wall inside the aorta is at higher temperature (T_h) compared with the outside wall (T_c). Heat transfer, $\dot{Q}(W)$, is in the direction of x and perpendicular to the plane of temperature difference, as shown in Fig. 3.1.

Heat transfer is a function of the higher and lower temperatures of the aorta wall, and aorta geometry and properties and is given by [1]:

$$\dot{Q} \propto \frac{(A)(\Delta T)}{\Delta x} \tag{3.1}$$

or

$$\dot{Q} = kA\frac{(T_h - T_c)}{\Delta x} = -kA\frac{(T_c - T_h)}{\Delta x} = -kA\frac{\Delta T}{\Delta x} \tag{3.2}$$

In Eq. (3.2), thermal conductivity $\left(k, \frac{W}{mK}\right)$ is transport property. Parameter A is the cross-sectional area (m^2) of the aorta and Δx is the aorta wall thickness (m). In the limiting case of $\Delta x \to 0$ Eq. (3.2) reduces to Fourier's law of conduction:

$$\dot{Q} = -kA\frac{dT}{dx} \tag{3.3}$$

where $\frac{dT}{dx}$ is the temperature gradient and must be negative based on the second law of thermodynamics. A more useful quantity to work with is heat flux, $q''\left(\frac{W}{m^2}\right)$, the heat transfer per unit area:

$$q'' = \frac{\dot{Q}}{A} \tag{3.4}$$

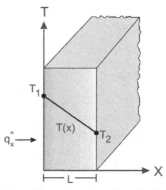

Figure 3.1 One-dimensional conduction heat transfer.

Separating the variables in Eq. (3.3), integrating from $x = 0$ and rearranging gives:

$$\dot{Q} = kA\frac{\Delta T}{L} \tag{3.5}$$

where L is the aorta wall thickness. For steady flow, one-dimensional heat conduction with no shaft work and no mass flow, the first law reduces to:

$$d\dot{Q} = \frac{d\dot{Q}(x)}{dx} = 0 \tag{3.6}$$

Combining Eqs. (3.3) and (3.6) gives:

$$\frac{d}{dx}\left(kA\frac{dT}{dx}\right) = 0 \tag{3.7}$$

If properties are assumed constant and by using the chain rule, the energy equation based on temperature is:

$$\frac{d^2 T}{dx^2} + \left(\frac{1}{A}\frac{dA}{dx}\right)\frac{dT}{dx} = 0 \tag{3.8}$$

Solving Eq. (3.8) provides the temperature field in a plane wall. Steady flow and one-dimensional heat transfer rate in cylindrical coordinate is:

$$\dot{Q} = -kA\frac{dT}{dr} \tag{3.9}$$

Separating the variables in Eq. (3.9), integrating from $r = 0$ and rearranging gives [1]:

$$\dot{Q} = (2\pi L)k\frac{\Delta T}{\ln(r_o/r_{in})} \tag{3.10}$$

where $A = 2\pi r L$, and r_o and r_{in} are the outside and inside wall radii.

Thermal Resistance Circuits: For steady one-dimensional flow with no generation of heat conduction equation, Eq. (3.5) can be rearranged as:

$$\dot{Q} = \frac{\Delta T}{L/KA} = \frac{\Delta T}{R_{cond}} \tag{3.11}$$

Conduction thermal resistance (R_{cond}) is represented by:

$$R_{cond} = \frac{L}{KA} \tag{3.12}$$

It is obvious that the thermal resistance R_{cond} increases as wall thickness (L) increases, and area (A) and K decrease. The concept of a thermal resistance circuit can be used for problems such as composite wall thickness (see Fig. 3.2).

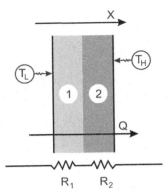

Figure 3.2 Heat transfer across a composite slab.

The heat transfer rate for composite wall is given by:

$$\dot{Q} = \frac{\Delta T}{\sum R_{cond}} = \frac{\Delta T}{R_1 + R_2} \tag{3.13}$$

3.1.2 Convective Heat Transfer

Convection is the energy transfer between two media, typically a surface and fluid that moves over the surface. In convective heat transfer heat is transferred by diffusion (conduction) and by bulk fluid motion (advection). An example of convection heat transfer is the flow of blood inside the blood vessels or air and water flow over the human skin. In convective heat transfer it is important to examine the fluid motion near the surface. Close to the wall there exists a thin layer called the "boundary layer" where fluid experiences velocity and temperature differences. Boundary layer thickness depends on flow Reynolds number, structure of the wall surface, pressure gradient, and Mach number. Outside this layer, temperature and velocity are uniform and identical to free stream temperature and velocity.

The rate of convection heat transfer $\left(\dot{Q}\right)$ from/to the surface is given by Newton's Law of Cooling as [1]:

$$\dot{Q} = hA(T_w - T_\infty) \tag{3.14}$$

The quantity h (W/m^{2}k) is called convective heat transfer coefficient and T_w and T_∞ are surface and fluid temperatures, respectively. For many situations of practical interest, the quantity h is known mainly through experiments. Integrating Eq. (3.14) over the entire surface leads to average convection coefficient $\left(\bar{h}\right)$ for the entire surface:

$$\bar{h} = \frac{1}{L} \int_{A_s} h dA_s \tag{3.15}$$

A thermal resistance is also defined for convection heat transfer from Eq. (3.16):

$$\dot{Q} = \frac{\Delta T}{1/hA} = \frac{\Delta T}{R_{conv}} \tag{3.16}$$

The convective thermal resistance (R_{conv}) is then:

$$R_{conv} = \frac{1}{hA} \tag{3.17}$$

Another important factor in convective heat transfer is friction coefficient (C_f), the characteristic of the fluid flow, which is:

$$C_f = \frac{\tau_w}{\rho_\infty u_\infty^2 / 2} \tag{3.18}$$

3.1.3 Radiation Heat Transfer

All bodies radiate energy in the form of photons. A photon is the smallest discrete amount of electromagnetic radiation (i.e., one quantum of electromagnetic energy is called a photon). Photons are massless and move in a random direction, with random phase and frequency. The origin of radiation is electromagnetic and is based on the Ampere law, the Faraday law, and the Lorentz force. Maxwell analytically showed the existence of electromagnetic wave. Electromagnetic waves transport energy at the speed of light in empty space and are characterized by their frequency (ν) and wavelength (λ) as follows:

$$\lambda = \frac{C}{\nu} \tag{3.19}$$

where C is the speed of light in the medium.

Electromagnetic waves appear in nature for over an unlimited range of wavelengths. Radiation with wavelength between 0.1 and 100 μm is in form of thermal radiation, and its transfer is called radiation heat transfer. Thermal radiation includes the entire visible and infrared as well as a portion of ultraviolet radiation.

All bodies at a temperature above absolute zero emit radiation in all directions over a wide range of wavelengths. The amount of emitted energy from a surface at a given wavelength depends on the material, condition, and temperature of the body. A surface is said to be diffuse if its surface properties are independent of direction and gray if its properties are independent of wavelength.

A black body is an ideal thermal radiator. It absorbs all incident radiation (absorptivity, $\alpha = 1$), regardless of wavelength and direction. It also emits the maximum radiation energy in all directions (diffuse emitter). The energy radiated per unit area is given by Joseph Stefan as [1]:

$$\dot{Q}_b'' = \sigma T^4 \tag{3.20}$$

Eq. (3.20) is called Stefan—Boltzmann law; σ is the Stefan—Boltzmann constant and is equal to $5.67 \times 10^{-8} \frac{W}{m^2 \cdot K^4}$, E_b is the blackbody emissive power, and T is the absolute temperature. Real bodies radiate less effectively than black bodies. The rate of real body radiation energy per unit area is defined by [1]:

$$\dot{Q}'' = \varepsilon \dot{Q}_b'' = \varepsilon \sigma T^4 \tag{3.21}$$

where ε is a property called the emittance. Values of emittance vary greatly for different materials. The emissivity of the human body is 0.97 for incident infrared radiation. The values are near unity for rough surfaces such as ceramics or oxidized metals, and roughly 0.02 for polished metals or silvered reflectors.

Radiation energy can be absorbed, reflected, or transmitted when it reaches a surface in human body. The sum of the absorbed, reflected, and transmitted fractions of radiation energy is equal to unity:

$$\alpha + \rho + \tau = 1 \tag{3.22}$$

where α is absorptivity (fraction of incident radiation that is absorbed), ρ is reflectivity (fraction of incident radiation that is reflected), and τ is transmissivity (fraction of incident radiation that is transmitted). Reflective energy may be either diffuse or specular (mirror-like). Diffuse reflections are independent of the incident radiation angle. For specular reflections, the reflection angle equals the angle of incidence.

3.1.4 Mass Transfer

Mass transfer is the net movement of mass from one location to another or travel of individual chemical species from high-concentration regions to low-concentration regions. Ordinary diffusion and convection are the two most important mechanisms in mass transfer phenomena. The most common example of mass transfer in the body is the purification of blood in the kidneys and liver or atherosclerotic disease in large arteries (coronary arteries) due to deposition of plaque on the wall of the arteries. Low-density lipoproteins (LDLs) (a main carrier of cholesterol) circulate within the blood stream and eventually accumulate in the artery walls and form fibrous plaques, also known as atherosclerotic patches. The deposition of plaque narrows the lumen and reduces the blood flow and leads to ischemia, heart failure, and cardiac death. Plaque deposition can also stiffen the vessel wall and can cause peripheral emboli (e.g., cerebral infarction, stroke).

Diffusion of particles in a gas mixture due to concentration gradients of molecular species is called diffusion phoresis. Diffusion is described by Fick's law, which relates the diffusion, J, to concentration, C, by:

$$J = -D\nabla C \tag{3.23}$$

where D is the diffusion coefficient and the negative sign indicates that diffusion occurs in the direction of decreasing concentration. ∇C is the concentration gradient and is as follows:

$$\nabla C = \hat{i}\frac{\partial C}{\partial x} + \hat{j}\frac{\partial C}{\partial y} + \hat{k}\frac{\partial C}{\partial z} \quad (3.24)$$

Thermal diffusion (the Soret effect) is the diffusion due to a temperature gradient and is usually considered when the temperature gradient is very large. Pressure diffusion is diffusion due to pressure gradient and negligible unless the pressure gradient is very large such as centrifugation process. Forced diffusion is the consequence of an external force (an electrical field) acting on a molecule.

The analysis of steady and incompressible blood flow, as a Newtonian fluid, in human coronary artery if decoupled from mass concentration field is determined by the continuity and Navier–Stokes equations as follows, respectively:

$$\nabla \cdot v = 0 \quad (3.25)$$

$$\rho(v\cdot\nabla)v = -\nabla P + \mu\nabla^2 v \quad (3.26)$$

LDLs are small spherical macromolecules that are transported passively by blood flow itself and voluntarily by molecular diffusion. The steady-state mass transport of LDL for blood flow in coronary artery is modeled by the convection–diffusion equation as:

$$v\cdot\nabla C - D\nabla^2 C = 0 \quad (3.27)$$

where C is the concentration of LDL, and D is the diffusivity of LDL. For low Reynold's number and spherical particle D is estimated by the Stones–Einstein equation:

$$D = \frac{k_B T}{6\pi\eta\, r} \quad (3.28)$$

where k_B is Boltzmann's constant ($1.38064852 \times 10^{-23}$ J/K), η is the dynamic viscosity, r is the radius of spherical particle, and T is absolute temperature. For blood with temperature around 37°C the diffusivity of LDL, using the Stokes–Einstein equation, is about 2×10^{-13} m^2/s [2].

In many mass-transfer operations in biological case studies the Sherwood number, an important dimensionless number, also known as the mass transfer Nusselt number, is used. The Sherwood number is the ratio of the convective mass transfer to the rate of diffusive mass transport and is given by:

$$Sh = \frac{h_m L_c}{D} \quad (3.29)$$

where h_m is the mass transfer coefficient, L_c is the characteristic length, and D is the mass diffusivity. The Nusselt and Sherwood numbers represent the effectiveness of heat and mass convection at the surface of a human body, respectively. For forced convection of blood flow inside a vessel, there exists an important analogy between the Nusselt number, which depends on the Reynolds and the Prandtl numbers, and the Sherwood number, which depends on the Reynolds and the Schmidt numbers (Sc). In other words, the Sherwood number can be obtained from the Nusselt number expression by simply replacing the Prandtl number by the Schmidt number, $Sh = f(Re, Sc)$. Schmidt number is the ratio of momentum diffusivity (ν) to mass diffusivity (D), $Sc = \nu/D$. For example, for fully developed turbulent blood flow in smooth vessel, the Nusselt and Sherwood numbers are as follows, respectively:

$$Nu_D = 0.023 Re_D^{0.8} Pr^{0.4} \quad Pr > 0.5 \tag{3.30}$$

$$Sh_D = 0.023 Re_D^{0.8} Sc^{0.4} \quad Sc > 0.5 \tag{3.31}$$

3.2 MICROSCALE HEAT TRANSFER

Microscale is referred to very small or microscopic scale items that are under 1 mm. Microscale flow is used in the study of the circulatory system, cell adhesion, and blood molecular transport. Microscale flows are central to biomedical applications that have emerged over the past few years. Microscale heat transfer is the transfer of thermal energy at the microscale level. There are two types of microscale heat transfer [3]: (1) thermal energy transfer related to small spaces, spaces that are sized between 1 and 200 μm, and (2) thermal energy related to small time domain with frequency up to several hundred hertz. For example, in human alveoli vessel that is sized around 175 μm in diameter there is only thermal energy transfer. On the other hand, in vessels that are sized less than 20 μm, such as arterioles and venules, not only thermal energy transfer occurs but also mass transfer takes places. However, in smaller sized human vessels such as the arterioles, venules, and capillaries (the smallest human vessels) mass transfer is the dominant phenomenon and thermal energy transfer is negligible.

Heat transfer and fluid flow in normal sized vessels are treated as continuous and continuum model is utilized. In the continuum model the general conservation of mass, momentum, and energy is considered. However, in microscale flow domain (such as the blood vessels) the continuum treatment of heat transfer and fluid flow breaks down because the vessel size is no longer enormously greater than the fluid molecular scale. Therefore the fluid must be considered in terms of collections of molecules using models such as the Boltzmann transport equation, lattice dynamics, molecular dynamics, and Monte Carlo.

In the human body there are several organs such as blood vessel, lungs, brain, and kidney that behave like microchannel. Different blood vessel sizes in the human body are shown in Fig. 3.3 [4].

Table 3.1 presents the channel classification.

Microchannels, which range between 10 and 200 μm, are influenced by the rarefaction effects for many gases and are described by the Knudsen number, Kn, as:

$$Kn = \frac{\lambda}{D} \tag{3.32}$$

where D is the vessel diameter and λ is the mean free path of the gas molecules and is calculated from:

$$\lambda = \frac{\mu\sqrt{\pi}}{\rho\sqrt{2RT}} \tag{3.33}$$

where R is the gas constant, μ is the viscosity, ρ is the density, and T is the absolute temperature in Kelvin. In the hydrodynamic regime, $Kn \ll 1$, the gas behaves as a continuum fluid on the length scale D, whereas in the Knudsen regime, $Kn \gg 1$, the continuum model is no longer valid. Table 3.2 gives the details of different flow regimes as a function of Knudsen number.

3.3 BIOHEAT TRANSFER

Bioheat transfer, as a subfield of biomedical and bioengineering, describes the transfer of thermal energy in biological systems. Heat transport in biological fields such as hyperthermia, organ storage, and restoration of biological tissue has attracted widespread attention. The transport of thermal energy in living tissue depends on temperature. Heat transfer in living tissue is a complex process and includes conduction, convection, and blood perfusion (such as delivery of the arterial blood to a capillary bed), cooling of human body by radiation, and metabolic heat generation. Metabolic heat generation results from a series of chemical reactions occurring in the living cells and blood perfusion occurs due to energy exchange between the living tissue and blood flow through small vessels in the living tissue. Of particular importance in bioheat transfer is the determination of the living tissue (e.g., skin, fat, muscle, bone, and blood) properties such as specific gravity, specific heat, and thermal conductivity.

Figure 3.3 The different blood vessel sizes in human body.

Table 3.1 The channel classification

Conventional channels	>3 mm
Minichannels	3 mm $\geq D >$ 200 μm
Microchannels	200 μm $\geq D >$ 10 μm
Transitional microchannels	10 μm $\geq D >$ 1 μm
Transitional nanochannels	1 μm $\geq D >$ 0.1 μm
Nanochannels	0.1 μm $\geq D$

D is the smallest channel dimension.

Table 3.2 Different flow regimes in function of Knudsen number (Kn)

Kn	Flow model	Method of calculation
$Kn < 10^{-3}$	Continuum flow	Navier–Stokes, classical no–slip BC
$10^{-3} < Kn < 10^{-1}$	Slip flow	Navier–Stokes, a velocity slip and a temperature jump, DSMC
$10^{-1} < Kn < 10$	Transition flow	Navier–Stokes is not valid, the intermolecular collision is considered. BTE, DSMC
$Kn > 10$	Free molecular flow	The occurrence of intermolecular collisions is negligible compared with the collisions between the gas molecules and the walls. BTE, DSMC

BC, boundary condition; BTE, Boltzmann transport equation; DSMC, direct simulation Monte Carlo.

3.3.1 Tissue Thermal Properties

In biological systems heat transfer is affected by many transient physiological and physical parameters such as vessel geometry, blood flow rates, and living tissue properties such as specific gravity, specific heat, and thermal conductivity. One of the techniques used to determine the key properties is the thermistor heating technique, a subdivision of thermal probe techniques [5–8]. Thermal properties of different living tissues and human parts as a function of temperature are listed in Table 3.3 [9,10].

The thermal conductivity (W/cm·°C) and thermal diffusivity (cm²/sec) of different items are determined as follows, respectively:

$$k = k_0 + k_1 T \tag{3.34}$$

$$\alpha = \alpha_0 + \alpha_1 T \tag{3.35}$$

Table 3.3 Thermal properties as a function of temperature

Tissue	K_0 m·W/cm·°C	K_1 m·W/cm·°C²	α_0 cm²/sec	α_1 cm²/sec·°C
Cerebral cortex	5.043	0.00296	0.001283	0.000050
Fat of spleen	3.431	−0.00254	0.001321	−0.000002
Liver	4.692	0.01161	0.001279	0.000036
Lung	3.080	0.02395	0.001071	0.000082
Lung	4.071	0.01176	0.001192	0.000031
Myocardium	4.925	0.01195	0.001289	0.000050
Pancreas	4.365	0.02844	0.001391	0.000084
Renal cortex	4.989	0.01288	0.001266	0.000055
Spleen	4.913	0.01300	0.001270	0.000047

Furthermore, the corresponding average thermal conductivity and thermal diffusivity are calculated by, respectively:

$$k = 4.574 + 0.01403\,T \tag{3.36}$$

$$\alpha = 0.001284 + 0.000053\,T \tag{3.37}$$

The thermal conductivity and thermal diffusivity of the aortic wall samples for 35 and 55°C are listed in Table 3.4 [10].

Thermal properties of tissue is a function of the concentration of its components [11,12].

$$K = 0.54 + 5.37 m_{water} \quad \text{for} \quad m_{water} > 0.2 \tag{3.38}$$

$$k = \rho \sum_n \frac{k_n m_n}{\rho_n} = \rho(6.28 m_{water} + 1.17 m_{protein} + 2.31 m_{fat}) \tag{3.39}$$

$$c = \sum_n c_n m_n = 4.2 m_{water} + 1.09 m_{protein} + 2.3 m_{fat} \tag{3.40}$$

Table 3.4 Thermal conductivity and thermal diffusivity of human aorta and atherosclerotic plaque

Tissue	Thermal conductivity (m·W/cm·°C)		Thermal diffusivity (× 1000 cm²/sec)	
	(at 35°C)	(at 55°C)	(at 35°C)	(at 55°C)
Normal aorta	4.76	5.03	1.27	1.33
Fatty plaque	4.84	4.97	1.28	1.32
Fibrous plaque	4.85	5.07	1.29	1.36
Calcified plaque	5.02	5.26	1.32	1.37

$$\rho = \frac{1}{\sum_n \frac{m_n}{\rho_n}} = k = \frac{1}{m_{water} + 0.649 m_{protein} + 1.227 m_{fat}} \tag{3.41}$$

where m_{water}, $m_{protein}$, and m_{fat} are the mass fraction of water, protein, and fat in the tissue, respectively, K is the thermal conductivity (m·W/cm·°C), c is the specific heat (J/g·°C), and ρ is the density (g/cm^3) of the tissue.

3.3.2 The Pennes Bioheat Equation

Temperature and heat transfer, which play an important role in biological systems such as human, are directly affected by blood flow in living tissues. Bioheat models such as Pennes equation are usually used to determine the transfer of heat in biological systems. Pennes proposed a model that includes the effects of blood perfusion, metabolism, and other source terms into the standard thermal diffusion equation (Eq. 3.5) within the tissue. The general Pennes bioheat equation that describes the temperature in a living tissue is given by [13]:

$$\rho C_P \frac{\partial T}{\partial t} = \nabla \cdot (k \nabla T) + Q_p + Q_m + Q_{others} \tag{3.42}$$

where ρ, C_p, T, and k are tissue density, tissue–specific heat, tissue temperature, and tissue thermal conductivity. Q_m is metabolic heat generation, the rate of energy deposition per unit volume, and is usually assumed to be homogeneously distributed throughout the tissue of interest. The metabolic heat generated in the body is dissipated to the environment through the skin and the lungs by convection and radiation as sensible heat. The total amount of sensible heat transfer from a human body exposed to air is calculated by:

$$\dot{Q}_{total} = \dot{Q}_{conduction} + \dot{Q}_{radiation} \tag{3.43}$$

$$\dot{Q} = h_c A(T_s - T_a) + \varepsilon \sigma A(T_s^4 - T_r^4) \tag{3.44}$$

Q_{others} is source terms such as hyperthermia. Q_p is the blood perfusion effect and is the thermal energy transfer between the blood and the tissue. Knowledge of the blood perfusion of tissues is important in that the flow of blood has a direct effect on the temperature distribution within living tissue. The blood perfusion is given by:

$$\dot{Q}_p = \rho C_p h_{perfusion}(T_{in} - T_{out}) \tag{3.45}$$

where T_{in} and T_{out} are the temperature of the blood upon entering and leaving the tissue via the arteriole-venule network. The blood perfusion effect is assumed to be homogeneous and isotropic, i.e., temperature is assumed to be linear.

3.3.3 Heat Transfer From Human Body

Human body continuously produces heat by burning the intake nutrients. Most body heat is generated in the deep organs, especially in the liver, brain, heart, and in contraction of skeletal muscles [14]. The body core temperature must be maintained around $37.1 \pm 1°C$ for a human to be healthy. Body temperature over $39°C$ can cause hyperthermia. Fig. 3.4 shows the body temperature in a hot and cold room [15].

The body core temperature as a function of age is also given in Table 3.5 [15].

The mean surface temperature of human body, T_s, is determined by [15]:

$$T_s = 0.07 T_{feet} + 0.32 T_{legs} + 0.18 T_{chest} + 0.17 T_{back} + 0.14 T_{arms} + 0.05 T_{hands}$$
$$+ 0.07 T_{head} \tag{3.46}$$

where T_{feet}, T_{legs}, T_{chest}, T_{back}, T_{arms}, T_{hand}, and T_{head} are temperatures of human feet, legs, chest, back, arms, hands, and head, respectively. To dissipate the excess heat from the human body the principles of engineering heat transfer are applied. The total amount

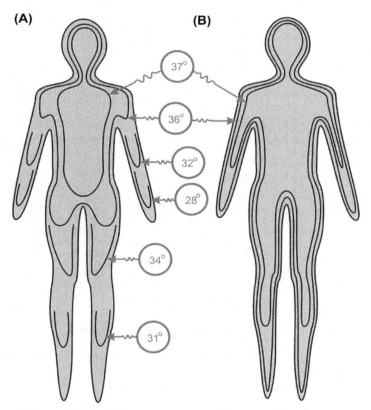

Figure 3.4 The body temperature in cold (A) and hot (B) room.

Table 3.5 The body core temperature for different ages

Age	Temperature (°C)
0—3 months	37.4
3—6 months	37.5
0.5—1 year	37.6
1—3 years	37.2
3—5 years	37
5—9 years	36.8
9—13 years	36.6
>13 years	36.5—37.2

energy produced in the human body is determined by the amount energy needed for basic body processes plus any amount of external work. The energy equation within the human body is given by:

$$Q_M - (W + Q_{cond} + Q_{conv} + Q_{rad} + Q_{evap} + Q_{rsp}) = \Delta E \tag{3.47}$$

where ΔE is the rate of energy storage in the body all in watt, Q_M is the amount of metabolic energy production, and W is the amount of external work. Q_{cond} is the amount of heat loss by conduction and is calculated by:

$$Q_{cond} = \frac{K_{cond}A_c}{L}(T_s - T_a)\Delta t \tag{3.48}$$

where T_s is the surface temperature of skin, T_a is the ambient temperature, L is cloth thickness, A_c is the covered body surface area, t is time, and K_{cond} is thermal conductivity of clothing material and is in Kcal/m·hr·°C. The suitable temperatures for insulation value, L/K_{cond}, of clothing are about 0.44 Clo for comfortable environment, about 0.1 Clo for hot environment, about 1 Clo for cold environment, and about 10 Clo for arctic environment.

Q_{conv} is the amount of surface heat loss by convection and is given by:

$$Q_{conv} = hA_c(T_s - T_a)\Delta t \tag{3.49}$$

where A_c is the covered body surface area. The different value of convective heat transfer coefficient, h (kcal/m²·hr·°C), is calculated by [15]:

$$h = 2.05(T_s - T_a)^{0.25} \quad \text{Free convection} \tag{3.50}$$

$$h = 5.6V^{0.67} \quad \text{Forced convection in standing and cross flow condition} \tag{3.51}$$

$$h = 2.54V^{0.72} \quad \text{Forced convection in standing and parallel flow condition} \tag{3.52}$$

where V (m/s) is the approach velocity.

Q_{rad} is the amount of surface heat loss by radiation and is calculated by:

$$Q_{rad} = \varepsilon\sigma A_c(T_s - T_a)\Delta t \qquad (3.53)$$

where σ, ε, and A_u are the Stefan–Boltzmann constant, the emissivity of the skin, and the exposed (uncovered) body area, respectively. The emissivity of the human body is about 0.97 for incident infrared radiation, 0.82 for visible light (dark skin), and about 0.65 for visible light (white skin) [15].

The amount of surface heat loss by evaporation, Q_{evap}, is obtained from two sources [15]: (1) the amount of heat loss by diffusion of water through the skin, Q_d, which is obtain from:

$$Q_d = 0.35A(P_s - P_a)\Delta t \qquad (3.54)$$

and (2) the amount of heat loss from sweating, Q_s, which is obtained from:

$$Q_s = h_{evap}A_w(P_s - P_a)\Delta t \qquad (3.55)$$

where A, A_w, and P_a are the body surface, the wetted surface area, and the partial pressure of water vapor in the ambient temperature. Also P_s (mmHg) is the vapor pressure of water at skin temperature and is calculated from $P_s = 1.92T_s - 25.3$. Forced convection evaporation transfer coefficient, h_{evap}, is obtained from $h_{evap} = 12.7V^{0.634}$.

Q_{rsp} is the amount of respiratory heat loss and is obtained from:

$$Q_{rsp} = Q_{lr} + Q_{sr} \qquad (3.56)$$

where the amount of latent heat loss, Q_{lr}, and the amount of sensible heat loss, Q_{sr}, are calculated as follows, respectively [15]:

$$Q_{lr} = \dot{m}_a h_{fg}(\omega_o - \omega_i) \qquad (3.57)$$

$$Q_{sr} = \dot{m}_a C_p(T_o - T_i) \qquad (3.58)$$

where $\dot{m}_a$ is the mass in kilograms of breathed air in and out per hour and is $\dot{m}_a = 0.006 \times$ metabolic rate, ω_o and ω_i are expired and inspired air water content (kg of water/kg of dry air), respectively, and h_{fg} is the water latent heat of vaporization at the expired air temperature.

Since the body does not operate with 100% efficiency, only a fraction of the metabolic rate is applied to work and the remainder shows up as heat. The mechanical efficiency associated with metabolic energy utilization is zero for most activities except when the person is performing external mechanical work, such as in walking upstairs, lifting something to a higher level, or cycling on an exercise machine. When external work is dissipated into heat in the human body, the mechanical efficiency is negative. An example of negative mechanical efficiency is walking downstairs. Storage of energy takes place whenever there is an imbalance of production and dissipation mechanisms. In many instances, such as astronauts in space suits or military personnel in chemical defense garments, energy storage is forced due to the lack of appropriate heat exchange with the environment.

3.4 APPLICATION OF MAGNETIC FIELD IN HYPERTHERMIA

Nanoparticles combined with magnetic fields are one of the most important research areas in the field of biomedical engineering such as hyperthermia and nanodrug delivery. Direct current (DC) magnetic and electromagnetic fields are often used for controlling nanoparticles and hyperthermia treatment, respectively. In this case study the effect of DC and alternating current (AC) magnetic fields as well as the electromagnetic fields on nano-biofluid (nanoparticles mixed with non-Newtonian blood) is presented. Fig. 3.5 shows the magnetic flux density distribution acting on a blood channel. The width (W) and length (L) of the channel are considered to be 0.001 and 0.02 m, respectively. The intensity and frequency of the AC magnetic field are 20 teslas and 5.85 MHz.

The schematic of the DC magnet and intensity distribution of a half-wave length dipole antenna in 1 GHz (electromagnetic fields) near the blood channel is shown in Fig. 3.6.

The half-wave length dipole antenna is used and is fed with Gaussian pulse [16]. The blood is modeled as non-Newtonian fluid by the power law model [17]. Blood density, specific heat capacity, permeability, and permittivity are 1055 kg/m^3, 4200 kg/m^3,

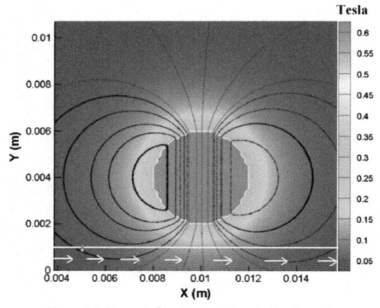

Figure 3.5 Magnetic flux density acting on blood vessel.

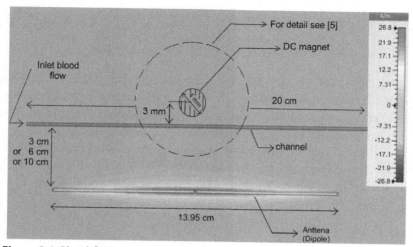

Figure 3.6 Blood flow exposed to direct current (DC) and electromagnetic fields.

29 N.A^{-2}, and 61 F/m, respectively [18,19]. The governing continuity, x and y momentum, and energy equations for the 2-D blood channel, represented in Fig. 3.5, with constant properties are as follows, respectively [18]:

$$\frac{\partial u}{\partial x} + \frac{\partial v}{\partial y} = 0 \tag{3.59}$$

$$\rho\left(u\frac{\partial u}{\partial x} + v\frac{\partial u}{\partial y}\right) = -\frac{\partial p}{\partial x} + \mu\left(\frac{\partial^2 u}{\partial x^2} + \frac{\partial^2 u}{\partial y^2}\right) + \left[\frac{1}{2}\mu_o\chi\frac{\partial}{\partial x}\left(\vec{H}\vec{H}\right)\right]C_v \tag{3.60}$$

$$\rho\left(u\frac{\partial v}{\partial x} + v\frac{\partial v}{\partial y}\right) = -\frac{\partial p}{\partial x} + \mu\left(\frac{\partial^2 v}{\partial x^2} + \frac{\partial^2 v}{\partial y^2}\right) + \left[\frac{1}{2}\mu_o\chi\frac{\partial}{\partial y}\left(\vec{H}\vec{H}\right)\right]C_v \tag{3.61}$$

$$\rho C\left(u\frac{\partial T}{\partial x} + v\frac{\partial T}{\partial y}\right) = k\left(\frac{\partial^2 T}{\partial x^2} + \frac{\partial^2 T}{\partial y^2}\right) + \dot{Q} \tag{3.62}$$

Volume concentration (C_v) and concentration equations are given by, respectively:

$$C_V = C_{V0}C \tag{3.63}$$

$$\vec{\nabla}\cdot\left(C\vec{v}_p\right) = \vec{\nabla}\cdot\left(D\vec{\nabla}C\right) \tag{3.64}$$

where C and C_{v0} are the dimensionless concentration and initial concentration of nanoparticles. The particle velocity (V_p) is determined by balancing the hydrodynamic and magnetic forces [18]. Diffusion coefficient (D) is obtained from Einstein equation [20]:

$$D = \frac{kT}{6\pi\mu r_P} \tag{3.65}$$

The source term $\left(\dot{Q}\right)$ in the energy equation is calculated from:

$$\dot{Q} = C\left(\dot{Q}_{hys} + \dot{Q}_{ec} + \dot{Q}_{res}\right) + \dot{Q}_{SAR} \tag{3.66}$$

where $\dot{Q}_{hys}$, $\dot{Q}_{ec}$, $\dot{Q}_{res}$, and $\dot{Q}_{SAR}$ are the hysteresis loss, the eddy current loss, residual loss, and the specific absorption rate (SAR) in the blood, respectively. For DC magnetic fields the hysteresis, eddy current, residual loss, and SAR are ignored. For AC magnetic field the hysteresis loss is ignored for superparamagnetic iron oxide nanoparticles [21].

The residual loss is due to different relaxation effects of magnetization in AC field. Neel and Brownian relaxation are two parts of residual loss. Neel relaxation τ_N is due to domain rotation in nanoparticles under AC magnetic field and is defined by [22]:

$$\tau_N = \tau_0 \exp\left(\frac{KV}{kT}\right) \tag{3.67}$$

Friction between particles and fluid τ_B causes energy loss and is given by [22]:

$$\tau_B = \frac{3\eta V_H}{k_B T} \tag{3.68}$$

The hydrodynamic volume V_H is related to magnetic field characteristic [21]. The effective relaxation time τ_{eff} and total relaxation loss (P) are defined as follows, respectively [23]:

$$\tau_{eff} = \frac{\tau_B \tau_N}{\tau_B + \tau_N} \tag{3.69}$$

$$P = \frac{\left(mH\omega\tau_{eff}\right)^2}{\left[2\tau_{eff}kTV\left(1 + \omega^2\tau_{eff}^2\right)\right]} \tag{3.70}$$

The DC magnetic field flux density is the dominant term because it is higher than the AC magnetic field intensity. Therefore the magnetic nanoparticles are fixed in one direction and do not rotate under AC magnetic field and as a result the relaxation loss due to AC magnetic field is small and is ignored.

The eddy current, the only energy source in magnetic nanoparticles induced by AC magnetic field, is calculated by [21]:

$$\dot{q} = \frac{\pi}{20}B_m^2 d^2 \sigma f^2 \times 10^{-16} \tag{3.71}$$

The SAR in blood is calculated as follows:

$$SAR = \frac{\sigma_{blood}}{\rho_{blood}}E_{rms}^2 \tag{3.72}$$

Root mean square of induced electric field in tissue (EF_{rms}) is obtained from:

$$EF_{rms} = \sqrt{\frac{EF_x^2 + EF_y^2 + EF_z^2}{3}} \qquad (3.73)$$

EF_x, EF_y, and EF_z are the induced electric field in tissue in different directions and are dependent on the external field (antenna exposure). Based on Figs. 3.5 and 3.6 the boundary conditions are as provided in Table 3.6.

Fig. 3.7 depicts the distribution of magnetic nanoparticles under the influence of the magnetic field when the blood inlet velocity is assumed to be 1 mm/s.

As shown the nanoparticles are absorbed with DC magnetic field near the upper wall. Velocity contour of the blood flow inside the channel under the influence magnetic field is shown in Fig. 3.8.

As shown in the figure, the magnetic nanoparticles shift toward the magnetic field. It causes the channel diameter to decrease. So the blood velocity is increased underneath the particles concentration.

Fig. 3.9 depicts the blood temperature contour in channel under the influence of magnetic field. The induced energy with magnetic field is dependent on magnetic nanoparticle concentration.

Also Fig. 3.10 depicts the blood temperature in the center line of channel under the influence of magnetic field.

According to Fig. 3.10 magnetic field has no effect on blood temperature. The electrical conductivity of nanoparticle superparamagnetic iron is much smaller than that of the bulk iron oxide. Moreover, the size of superparamagnetic iron oxide is so small. According to these reasons eddy current could not change the blood temperature in the channel under influence of magnetic field.

Fig. 3.11 depicts energy per volume inducing in blood flow under influence antenna wave, which is located at 3 cm below the channel.

Table 3.6 Boundary condition for Figs. 3.5 and 3.6

Boundary	Conditions
Channel inlet	Concentration is unit, $C_{inlet} = 1$
Channel upper wall	Velocity is zero
Channel lower wall	Velocity is zero
Channel outlet	Neumann boundary condition
Wall temperature	315.14 K
Blood temperature at inlet	310.15 K

Figure 3.7 Concentration distribution of nanoparticles inside the channel under the influence of magnetic field.

Figure 3.8 Velocity contour of the blood flow inside the channel under the influence of magnetic field.

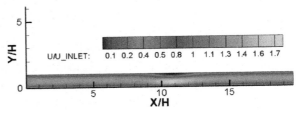

Figure 3.9 Blood temperature contour under influence of magnetic field.

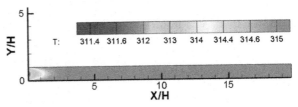

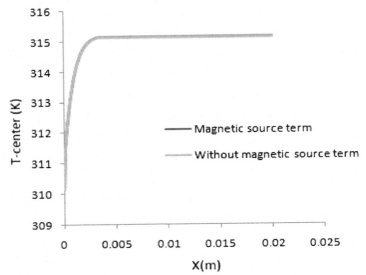

Figure 3.10 Comparison of blood temperatures in central line of channel with magnetic field and without magnetic field.

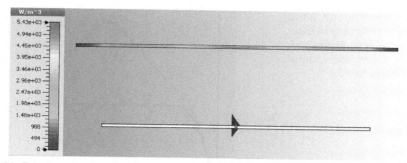

Figure 3.11 Energy per volume contour inducing in blood for distance 3 cm between antenna and channel.

Because of wave's intensity, the amount of energy absorption at the central portion of the channel is higher than the other part.

Fig. 3.12 shows the temperature distribution inside the channel under the influence of antenna exposure when the distance between the channel and antenna is 3 cm. The blood inlet and wall temperatures are 310.15 and 315.15 K, respectively. In addition, the blood inlet velocity is 0.001 mm/s.

As shown blood temperature at the center of channel is increased up to 5 degrees.

Fig. 3.13 depicts comparison of blood temperatures at center line of channel in x direction for two different case studies. In one case temperature is calculated when the antenna exposure is applied 3 cm below the channel. In addition, the blood temperature in the channel is obtained without the antenna.

The blood temperature contours when the channel is located at distance 3, 6, and 10 cm from antenna are shown in Fig. 3.14.

The blood inlet and wall temperatures are 310.15 and 315.15 K, respectively. Also the blood inlet velocity is 0.001 mm/s. Because of increasing distance from the antenna, the SAR in blood is decreased and the blood temperature is decreased too.

Fig. 3.15 depicts the blood temperature at central line of the channel when the channel is located at distance 3, 6, and 10 cm from the antenna.

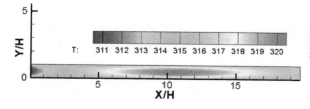

Figure 3.12 Temperature contour of blood flow inside the channel under the influence antenna exposure (3 cm gap space between antenna and channel).

Figure 3.13 Comparison of blood temperatures at center line of channel when the antenna exposure is applied 3 cm below the channel and without the antenna.

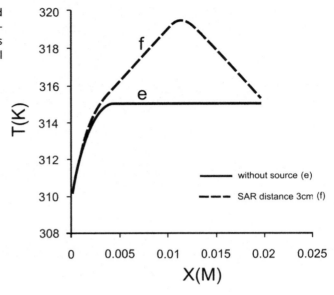

Figure 3.14 The blood temperature distribution inside the channel at different gap spaces between antenna and channel (A) 3 cm, (B) 6 cm, (C) 10 cm.

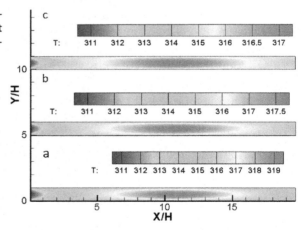

Fig. 3.16 shows the maximum blood SAR in channel at different diameters under the influence of antenna exposure when the distance between the channel and antenna is 3 cm.

As expected, increasing the diameter increased the blood SAR and temperature in blood.

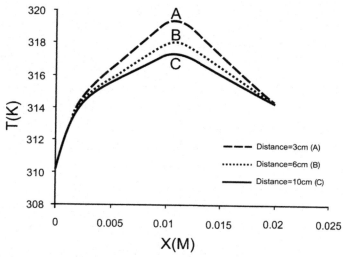

Figure 3.15 Blood temperature in the central line of the channel at different gap spaces between antenna and channel (A) 3 cm, (B) 6 cm, (C) 10 cm.

Results show that increase in blood temperature does not influence the nanoparticle concentration as well as the velocities and pressure distribution of blood inside the channel. In addition, the wave intensity increases the amount of nano–biofluid energy absorption at the central portion of the channel compared with the other part of the channel. The SAR and temperature of nano-biofluid are decreased by increasing the channel distance from the antenna or decreasing the channel diameter. Finally, it is shown that increasing the temperature of nano-biofluid under the influence of electromagnetic fields is useful for hyperthermia as well as drug release for drug delivery purposes.

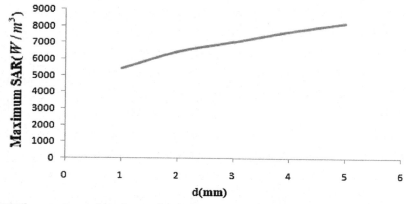

Figure 3.16 The maximum blood specific absorption rate (SAR) in channel at different diameters (with 3 cm gap space).

3.5 APPLICATION OF ULTRASONIC WAVE

High intensity focused ultrasound (HIFU) is a new field applied to treat solid malignant tumors in different parts of the body [24,25]. Another application of HIFU includes small blood vessel occlusion [26]. During focused therapy, the absorbed ultrasound energy in targeted tissue is transformed into thermal energy and the tissue temperature is increased. Using the Westervelt equation and considering diffraction, absorption, nonlinear propagation, and relaxation, the thermal effects of applying a HIFU source is determined by solving the following coupled system of partial differential equations as given below [27]:

$$\nabla^2 p - \frac{1}{c_0^2}\frac{\partial^2 p}{\partial t^2} + \frac{\delta}{c_0^4}\frac{\partial^3 p}{\partial t^3} + \frac{\beta}{\rho_0 c_0^4}\frac{\partial^2 p^2}{\partial t^2} + \sum_i P_i = 0 \tag{3.74}$$

$$\left(1 + \tau_i\frac{\partial}{\partial t}\right)P_i = \frac{2}{C_0^3}c_i\tau_i\frac{\partial^3 p}{\partial t^3} \tag{3.75}$$

where p is the sound pressure, β is the nonlinearity coefficient, δ is the sound diffusivity, τ_i is the relaxation time, and c_i is the sound speed increment. In Eq. (3.71) the first two terms represent the linear wave propagation and the third term calculates the thermal conduction and fluid viscosity losses. The acoustic nonlinearity effect is determined from the fourth term and the final term models the inevitable relaxation processes. The power of HIFU per unit volume (q_{HIFU}) is calculated by:

$$q_{HIFU} = \sum_{n=1}^{\infty} 2\alpha(nf_0)I_n \tag{3.76}$$

where n is the number of harmonics, α is the absorption coefficient, I_n is the intensity of nth harmonic, and f_0 is the main frequency of ultrasonic waves. For example, the ultrasonic intensity of linear Westervelt equation is:

$$I_L = \frac{p^2}{2\rho c_0} \tag{3.77}$$

The human body blood vessels and tissues are affected while applying HIFU to the body. In tissues without large blood vessels, Pennes bioheat equation may be applied as below [28]:

$$\rho_t c_t \frac{\partial T}{\partial t} = k_t \nabla^2 T - \omega_b c_b(T - T_\infty) + q \tag{3.78}$$

In the above equation, ρ is the density, c is the specific heat coefficient, k is the thermal conductivity, and w_b is the perfusion rate. For blood media subscripts b is used.

Wherever large blood vessels exist or heat convection of blood flow cannot be neglected, energy equation converts to:

$$\rho_t c_t \frac{\partial T}{\partial t} = k_t \nabla^2 T - \rho_b c_b \overrightarrow{u} \cdot \nabla T + q \tag{3.79}$$

where u is blood flow velocity.

REFERENCES

[1] J.P. Holman, Heat Transfer, sixth ed., McGraw-Hill Book, 1st printing, 1986.
[2] D.W. Tank, W. Fredericks, L. Barak, W. Webb, Electric field-induced redistribution and postfield relaxation of low density lipoprotein, The Journal of Cell Biology 101 (1985) 148−157.
[3] P. Tabeling, Introduction to Microfluidics (S. Chen, Trans.), Oxford University Press, 2005.
[4] S. Kandlikar, S. Garimella, D. LI, S. Colin, M.R. King, Heat Transfer and Fluid Flow in Minichannels and Microchannels, Elsevier, 2004.
[5] J.C. Chato, A method for the measurement of thermal properties of biologic materials, in: Symposium on Thermal Problems in Biotechnology, ASME, New York, LCN068-58741, 1968, pp. 16−25.
[6] T.A. Balasubramaniam, H.F. Bowman, Thermal conductivity and thermal diffusivity of biomaterials: a simultaneous measurement technique, Journal of Biomechanical Engineering 99 (1977) 148−154.
[7] J.W. Valvano, et al., An isolated rat liver model for the evaluation of thermal techniques to measure perfusion, Journal of Biomechanical Engineering 106 (1984) 187−191.
[8] H. Arkin, et al., Thermal pulse decay method for simultaneous measurement of local thermal conductivity and blood perfusion: a theoretical analysis, Journal of Biomechanical Engineering 108 (1986) 208−214.
[9] J.W. Valvano, et al., Thermal conductivity and diffusivity of biomaterials measured with self-heated thermistors, International Journal of Thermophysics 6 (1985) 301−311.
[10] J.W. Valvano, B. Chitsabesan, Thermal conductivity and diffusivity of arterial wall and atherosclerotic plaque, Lasers in the Life Sciences 1 (1987) 219−229.
[11] K.E. Spells, The thermal conductivities of some biological fluids, Physics in Medicine & Biology 5 (1960) 139−153.
[12] T.E. Cooper, G.J. Trezck, Correlation of thermal properties of some human tissues with water content, Aerospace Medicine 42 (1971) 24−27.
[13] H.H. Pennes, Analysis of tissue and arterial blood temperature in the resting human forearm, Journal of Applied Physics 1 (1948) 93−122.
[14] A.C. Guyton, J.E. Hall, Textbook of Medical Physiology, eleventh ed., Elsevier Saunders, Philadelphia, 2006, p. 890.
[15] S. Najarian, A. Abouei Mehrizi, Heat and Mass Transfer in Biological Systems, Jahad Daneshgahi. Amir Kabir Branch, October 2006.
[16] M.R. Mohammadi, Analysis of Blood Flow and Heat Transfer under the Effect of Magnetic Field, K.N. Toosi University of Technology, 2014 (Master of Science Thesis).
[17] A. Shahidian, et al., Flow analysis of non-Newtonian blood in a magnetohydrodynamic pump, IEEE Transactions on Magnetics and Engineering 45 (6) (2009).
[18] M.R. Habibi, M. Ghasemi, Numerical study of magnetic nanoparticles concentration in biofluid (blood) under the influence of high gradient magnetic field, Journal of Magnetism and Magnetic Materials 323 (2011) 32−38.
[19] M. Ghassemi, et al., A new effective thermal conductivity model for a bio-nanofluid (blood with nanoparticle Al2O3), International Communications in Heat and Mass Transfer 37 (2010) 929−934.
[20] M.R. Habibi, et al., Analysis of high gradient magnetic field effects on distribution of nanoparticles injected into pulsatile blood stream, Journal of Magnetism and Magnetic Materials 324 (2012) 1473−1482.

[21]   A. Shahidian, M.R. Mohammadi, M. Ghassemi, Numerical simulation of blood flow mixed with magnetic nanoparticles under the influence of AC and DC magnetic field, Iranian Journal of Mechanical Engineering 15 (1) (March 2014).

[22]   Y. Zhai, Y. Zhang, Magnetic induction heating of nano sized ferrite particle, in: S.Å. 'raw Grundas (Ed.), Advances in Induction and Microwave Heating of Mineral and Organic Materials, In Tech, 2011, pp. 484–500.

[23]   R. Ivkovet, Application of high amplitude alternating magnetic fields for heat induction of nanoparticles localized in cancer, Clinical Cancer Research 11 (2005) 7093s.

[24]   Y.F. Zhou, High intensity focused ultrasound in clinical tumor ablation, World Journal of Clinical Oncology 2 (2011) 8–27.

[25]   T.A. Leslie, J.E. Kennedy, High intensity focused ultrasound in the treatment of abdominal and gynaecological diseases, International Journal of Hyperthermia 23 (2007) 173–182.

[26]   K. Hynynen, V. Colucci, A. Chung, F. Jolesz, Noninvasive arterial occlusion using MRI-guided focused ultrasound, Ultrasound in Medicine and Biology 22 (1996) 1071–1077.

[27]   M.A. Solovchuk, M. Thiriet, T.W.H. Sheu, Computational Study for Investigating Acoustic Streaming and Heating during Acoustic Hemostasis, November 24, 2014 arXiv:1411.6343v1 [physics.flu-dyn].

[28]   M.A. Solovchuk, T.W.H. Sheu, M. Thiriet, W.-L. Lin, On a computational study for investigating acoustic streaming and heating during focused ultrasound ablation of liver tumor, Applied Thermal Engineering 56 (2013) 62–67.

CHAPTER 4

Fluid Mechanics

4.1 FUNDAMENTAL CONCEPTS

The distinguished characteristic of fluids compared with solids is related to the amount of deformation rate. Fluids show a continuous deformation when they get exposed to a shear force. However, different fluids reveal different rates of deformation when a specified shear force acts on them. Consider a fluid between two very wide parallel plates, as shown in Fig. 4.1. According to Newton's law of viscosity, for common fluids such as water, oil, and gasoline, the ratio of shear stress existing in fluid layers to the velocity gradient is equal to a property of fluid named dynamic viscosity (for more details, readers are referred to Ref. [1]):

$$\frac{\tau}{du/dy} = \mu \tag{4.1}$$

where τ is the shear stress, μ is the dynamic viscosity, and du/dy is the velocity gradient.

Such fluids, for which shear stress linearly varies with the velocity gradient, are categorized as Newtonian fluids. On the other hand, for non-Newtonian fluids, shear stress does not vary linearly with velocity gradient.

In the following section first a fundamental understanding of blood, as an important biofluid, is introduced. Then the properties of blood vessel are discussed.

4.1.1 Hematology and Blood Rheology

Hematology is the study of the nature, function, and diseases of the blood as well as the blood-forming organs. On the other hand, rheology is the study of the matter deformation and its flow.

Blood contains antibodies, nutrients, oxygen, and much more to help the body work [1]. Blood, as an important human body fluid, delivers nutrients and oxygen to the cells

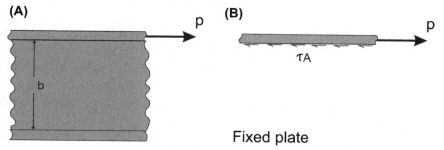

Figure 4.1 (A) Fluid located between two very wide parallel plates; (B) free body diagram.

Nano and Bio Heat Transfer and Fluid Flow
ISBN 978-0-12-803779-9, http://dx.doi.org/10.1016/B978-0-12-803779-9.00004-2

and in return transports metabolic waste products away from them [1]. When blood reaches the lungs it exchanges the gases; carbon dioxide diffuses out of the blood into the pulmonary alveoli while oxygen diffuses into the blood. The oxygenated blood is then pumped into the left-hand side of the heart in the pulmonary vein and enters the left atrium. From here it passes through the mitral valve, through the ventricle, and throughout the body through the aorta.

Human blood is similar to mammal's blood, although the precise details concerning cell numbers, size, protein structure, and so on vary somewhat between species. Blood carries out many important functions within the body. Among many, it supplies oxygen to the tissues (oxygen bound to the hemoglobin, which is carried by red cells); supplies nutrients such as glucose, amino acids, and fatty acids [dissolved in the blood or bound to plasma proteins (e.g., blood lipids)]; removes waste such as carbon dioxide, urea, and lactic acid; and regulates body pH and core body temperature.

Blood accounts for 7% of the human body weight [2], with an average density of approximately 1060 kg/m^3, very close to pure water's density of 1000 kg/m^3 [3]. The average adult has a blood volume of roughly 5 L [2]. In general blood is composed of plasma and several kinds of cells (also called corpuscles or "formed elements"), that is, erythrocytes (red blood cells), leukocytes (white blood cells), and thrombocytes (platelets). By volume, the red blood cells constitute about 45% of whole blood, the plasma about 54.3%, and white cells about 0.7% [4]. Fig. 4.2 shows the elements of blood.

Blood pH is regulated between the narrow range of 7.35–7.45; blood pH below 7.35 is too acidic and blood pH above 7.45 is too basic [5]. Blood pH as well as oxygen and carbon dioxide partial pressure and bicarbonate (HCO_3^-) are carefully regulated by a number of homeostatic mechanisms, which exert their influence principally through the respiratory system and the urinary system to control the acid–base balance and respiration.

Blood consists of 40%–45% formed elements, including red blood cells or erythrocytes, white blood cells or leukocytes, and platelets or thrombocytes. Erythrocytes primarily involve oxygen and carbon dioxide transport. Leukocytes primarily involve phagocytosis and immune responses, whereas thrombocytes involve blood clotting. In addition to the formed elements in blood, 55%–60% of blood by volume consists of plasma [6]. Plasma is a transparent and amber-colored liquid in which the cellular components of blood are suspended. Plasma contains proteins, electrolytes, hormones, and nutrients. Serum is blood plasma from which clotting factors are removed (Tables 4.1 and 4.2) [7].

The thermophysical properties of blood depend on different parameters such as age, temperature, and hematocrit. However, in living organisms in general, and in large mammals in particular, general properties do not alter significantly because all mentioned parameters are regulated [8].

Blood viscosity is determined by several factors such as the viscosity of plasma, hematocrit level, blood cell distribution, and mechanical properties of blood cells [6,7]. Blood viscosity is

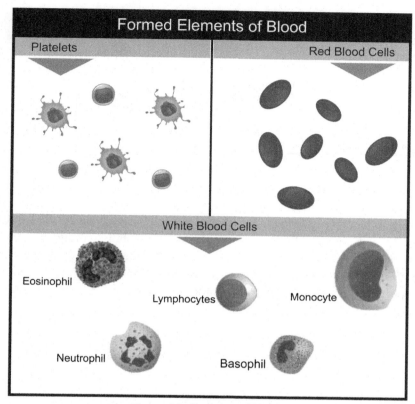

Figure 4.2 Elements of blood.

Table 4.1 Plasma characteristics [7]

Plasma component	Component	Molecular weight (dalton)	Density (*g/dl*)
Plasma	Water 91% Protein 7%		
	Albumin	69,000	4.5
	Fibrinogen	340,000	0.3
	Immunoglobulins	140,000	2.5
	Prothrombin	68,700	0.015
	Salt	Soluble in water	
Other	Includes vitamin, lipid, sugar, etc.		

also affected by the applied deformation forces, extensional as well as shearing, and the ambient physical conditions. Although plasma is essentially a Newtonian fluid, the blood as a whole behaves as a non–Newtonian fluid showing all signs of non–Newtonian rheology, which includes deformation rate dependency, viscoelasticity, yield stress, and thixotropy. Most non–Newtonian effects originate from the red blood cells due to their high

Table 4.2 Blood cell characteristics [30]

Blood component		Per microliter	Size (μm)	%
Red blood cell		$4.1-5.1 \times 10^6$	$7-8$	97
White blood cell (WBC)	Neutrophils	62% of WBC	$10-12$	2
$4-10 \times 10^3$	Lymphocytes	30% of WBC	$6-14$	
	Eosinophils	2.3% of WBC	$-$	
	Monocytes	5.3% of WBC	$15-20$	
	Basophils	0.4% of WBC	$-$	
Platelet		$1.5-4.5 \times 10^5$	3	1

concentration and distinguished mechanical properties such as elasticity and ability to aggregate, forming three-dimensional structures at low deformation rates [1]. Another important factor that affects blood viscosity is the Fahraeus–Lindqvist effect where the viscosity of blood changes with the diameter of the vessel it travels through as shown in Fig. 4.3.

This is especially true when the vessel (tubes) diameter is less than approximately 1 mm. In this case blood behaves as a non-Newtonian fluid [2].

An important equation that estimates blood viscosity at various temperatures and hematocrits is Einstein's equation, which is [2]:

$$\mu = \mu_{plasma}\left(\frac{1}{1 - \xi\psi}\right) \tag{4.2}$$

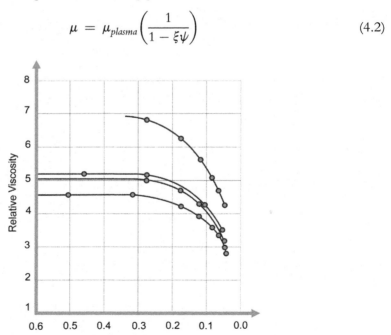

Figure 4.3 Viscosity as a function of diameter. *(Reproduced from the original 1931 article L. Waite, J. Fine, Applied Biofluid Mechanics, Mc-Graw Hill, 2007, http://dx.doi.org/10.1036/0071472177.)*

Table 4.3 Thermophysical properties of blood [3]

Blood thermophysical properties	Red blood cell	Plasma	Blood
Density (kg/m^3)	1125	1025	1060
Viscosity (cp)	6	1.2	3
Specific heat capacity (J/kg °C)	–	–	3750
Thermal conductivity (W/m K)	0.45–0.482	0.57	0.492
Electrical conductivity (S/m)	–	2	0.667

where μ is the whole blood viscosity, μ_{plasma} is the plasma viscosity, ψ is the hematocrit, and ξ is defined in the following equation [2]:

$$\xi = 0.076 \times \exp\left\{2.49\psi + \frac{1107}{T}\exp(-1.69\psi)\right\} \tag{4.3}$$

Table 4.3 shows the thermophysical properties of blood.

4.1.2 Structure of Blood Vessel

The cardiovascular system is made up of the heart (as a central pump) and the blood vessels. It is divided into three subsystems: the systemic circulation, the pulmonary circulation, and the coronary circulation [9]. Fig. 4.4 depicts the systemic circulation and the pulmonary circulation.

The heart supplies the force that causes the blood to flow through the circulatory system. It is a four-chambered pump with two upper chambers, known as atria, and two lower chambers, known as ventricles; see Fig. 4.5. Check valves between the chambers ensure that the blood moves in only one direction, enable the pressure in the lungs, and restrict the blood from flowing backward from the aorta toward the lungs.

Blood enters the right atrium from the vena cava and then is pumped into the right ventricle. From the right ventricle, blood is pumped downstream through the pulmonary artery to the lungs where it is enriched with oxygen and gives up carbon dioxide. On the left side of the heart, oxygen–enriched blood enters the left atrium from the pulmonary vein. When the left atrium contracts, it pumps blood into the left ventricle. When the left ventricle contracts, it pumps blood at a relatively high pressure, ejecting it from the left ventricle into the aorta.

The left side of the heart supplies oxygenated blood to the aorta at a relatively high pressure. The path of blood flow through the circulatory system and the blood pressures at various points along the way determine how tissue is perfused with oxygen. The oxygenated blood initially flows into the smaller arteries and finally into systemic capillaries where oxygen is supplied to the surrounding tissues. At the same time, blood picks up carbon dioxide from the same tissue and continues flowing into the veins. Eventually,

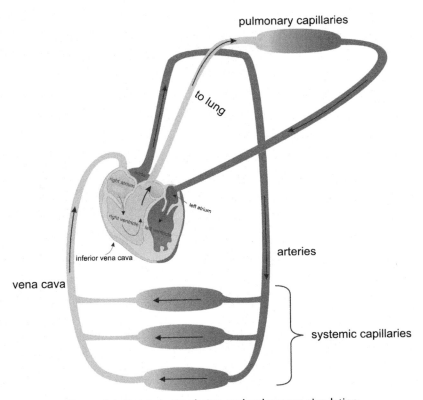

Figure 4.4 Systemic circulation and pulmonary circulation.

the blood returns to the vena cava. From the vena cava, deoxygenated blood flows into the right side of the heart. From there, the still deoxygenated blood flows into the pulmonary artery. The pulmonary artery supplies blood to the lungs where carbon dioxide is exchanged with oxygen. The blood, which has been enriched with oxygen, flows from the lungs through the pulmonary veins and back to the left heart. It is interesting to note that blood flowing through the pulmonary artery is deoxygenated and blood flowing through the pulmonary vein is oxygenated. A more appropriate distinction between arteries and veins is that arteries carry blood at a relatively higher pressure than the pressure within the corresponding veins.

The systemic circulation is the subsystem supplied by the aorta that feeds the systemic capillaries. The pulmonary circulation is the subsystem supplied by the pulmonary artery that feeds the pulmonary capillaries. The coronary circulation is the specialized blood supplies that perfuse cardiac muscle. The cardiovascular system provides oxygen and nutrients to body tissues, removes carbon dioxide and other wastes from the body, and regulates the body temperature.

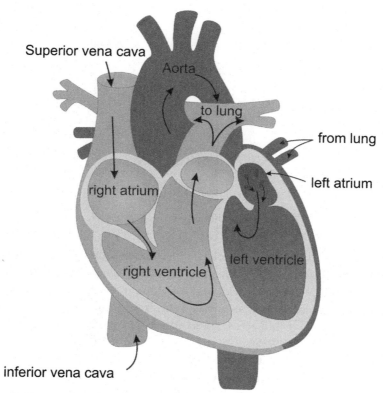

Figure 4.5 Four chambers of the heart, namely, the right atrium, right ventricle, left atrium, and left ventricle.

4.1.2.1 Arteries

Arteries are vessels with high blood pressure that carry blood away from the heart through smaller arteries, to arterioles, and further to the level of capillaries. Most arteries carry oxygenated blood except the pulmonary and the umbilical arteries. The effective arterial blood volume is that extracellular fluid which fills the arterial system.

In general, arteries are composed of three layers: the tunica intima or the innermost layer, the tunica media or the middle layer, and the tunica externa (the outermost layer of the artery); see Fig. 4.6 [2].

The lumen is inside the artery where the blood flows. The lumen is lined with the endothelium, which consists of simple squamous epithelial cells, to form an interface between the blood and the vessel wall [2]. The epithelial cells are in close contact and form a slick layer to prevent the interaction between the blood cells and the vessel wall as blood moves through the vessel lumen. The endothelium plays an important role in the mechanics of blood flow, blood clotting, and leukocyte adhesion. Also, it performs

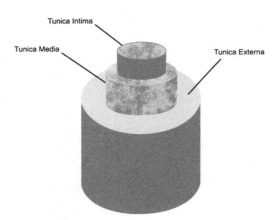

Figure 4.6 The three layers of an artery [2].

many other functions, such as the secretion of vasoactive substances and the contraction and relaxation of vascular smooth muscle. The innermost layer of an artery, also known as the tunica intima, is one cell layer thick and it is composed of endothelial cells.

The tunica media, also known as middle coat, is smooth muscle cells in the middle of the blood vessel. It is live and active and contracts or expands to change the diameter of the vessel and change the blood flow rate (the vasoconstriction and vasodilation). The tunica media also consists of elastic tissue or fiber that is passive and does not have significant metabolic activity. It not only supports the blood vessel but also allows an expanded blood vessel to retract when the pressure is removed. The tunica media is absent in capillaries. The tunica externa, also known as tunica adventitia, is composed of connective passive elastic tissue similar to the tunica media tissue and is absent in capillaries [11].

Arteries, by their functional characteristics, are divided into elastic arteries, muscular arteries, and arterioles. Among the three, the elastic arteries have the largest diameter and store additional volume of blood when subjected to higher pressure. Aorta is a good example of an elastic artery. Elastic arteries have very thick tunica media compared with other arteries and contain a large amount of the elastic fiber elastin.

Most arteries are muscular arteries, which are the intermediate-sized arteries [11]. In muscular arteries, the tunica media is composed almost entirely of smooth muscle. Functionally, muscular arteries can change diameter to influence flow by vasoconstriction and vasodilation.

Arterioles, arteries with diameter less than 0.5 mm, are composed entirely of smooth muscle cells. Arterioles do not possess much of a tunica externa. Approximately 70% of pressure drop between the heart and the veins occurs in the small arteries and arterioles.

Blood flows from the heart through the arteries and arterioles into capillaries. After perfusion of the tissues, capillaries join and widen to become venules, which in turn widen to become veins. Veins return the blood back into the heart through the great veins.

4.1.2.2 Veins

Veins as opposed to arteries are vessels that carry blood to the heart in the circulatory system. All veins, except for the pulmonary and umbilical, carry deoxygenated blood from the tissues back to the heart. Pulmonary and umbilical veins carry oxygenated blood to the heart. Veins are less muscular than arteries and are often closer to the skin. There are valves in most veins to prevent regurgitation (reverse blood flow). Veins conduct the blood from the capillaries back to the heart on the lower pressure side of the cardiovascular system.

In general, veins are essentially tubes that collapse when their lumens are not filled with blood. The thick outermost layer of a vein is made of connective tissue and is known as tunica adventitia or tunica externa [12]. The tunica media, the middle layer band of smooth muscle, is thin and like veins do not function primarily in a contractile manner [11]. The interior is lined with endothelial cells and is called tunica intima. The precise location of veins varies much more from person to person compared with arteries [12]. Veins often display a lot of anatomical variation compared with arteries within a species and between species.

4.1.2.3 Capillaries

Capillaries are the most common and thinnest of the blood and lymph vessels. They carry blood to the cells of the body tissues. The walls of capillaries consist of only a thin layer of endothelium. The thickness of the endothelial lining between the blood and the tissues is minimal (one cell layer thick) and it acts as a filter and separates the blood cells from other chemicals. It causes the blood cells to stay inside the vessels while allowing the rest of chemicals to diffuse into or out of tissues.

Capillaries usually function in a capillary bed and are divided into true capillaries and metarterioles. True capillaries branch from arterioles and provide exchange between cells and the blood. Metarterioles are short vessels and directly connect the arterioles and venules. In most cases, the true capillaries oxygenated blood going from the terminal branches (about $10-50$ μm in diameter) of the arterioles into the metarterioles ($10-20$ μm in diameter) and eventually merge with the noncontractile postcapillary ($8-30$ μm in diameter) and collecting venules ($10-50$ μm).

Lymphatic capillaries are slightly larger in diameter than blood capillaries, and have closed ends (unlike the loop structure of blood capillaries). This structure permits interstitial fluid to flow into but not out of them. Lymph capillaries have a greater internal oncotic pressure than blood capillaries, due to the greater concentration of plasma proteins in the lymph. Fig. 4.7 shows the three main types of blood capillaries.

4.1.2.4 Continuous Capillaries

These capillaries are continuous in the sense that the endothelial cells provide an uninterrupted lining. They only allow smaller molecules, such as water and ions, to pass through their intercellular clefts. However lipid-soluble molecules diffuse through the endothelial cell membranes along concentration gradients passively. Tight junction

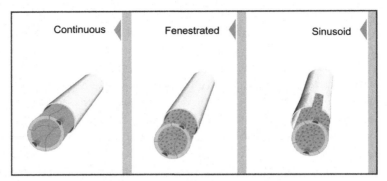

Figure 4.7 The three main types of blood capillaries [4].

capillaries are cells whose membranes join together forming a virtually impermeable barrier to fluid. They are those that are found in skeletal muscles, fingers, gonads, and skin and those found in the central nervous system.

4.1.2.5 Fenestrated Capillaries

These are capillaries that have pores with 60–80 nm diameter inside the endothelial cells. They are spanned by a diaphragm of radially oriented fibrils, which allows small molecules and limited amounts of protein to diffuse [13,14]. In renal glomerulus there are cells with no diaphragms, also known as podocyte foot processes or pedicels. They have slit pores with a function analogous to the diaphragm of the capillaries. Both blood vessels have continuous basal laminae and are primarily located in the endocrine glands, intestines, pancreas, and the glomeruli of the kidney.

4.1.2.6 Sinusoidal Capillaries

Sinusoidal capillaries, also known as discontinuous capillaries, are a special type of open–pore capillary with 30- to 40-μm diameter openings in the endothelium [13]. They are aided by a discontinuous basal lamina and allow red and white blood cells (7.5–25 μm diameter) and various serum proteins to pass. Lack of pinocytotic vesicles in sinusoidal capillaries permits transfer of blood between endothelial cells, and hence across the membrane. Sinusoidal blood vessels are primarily located in the bone marrow, lymph nodes, and adrenal glands. Some sinusoids are distinctive in that they do not have tight junctions between cells. For this reason they are called discontinuous sinusoidal capillaries. They are present in the liver and spleen, where greater movement of cells and materials is necessary.

4.1.3 Governing Equations

To have a comprehensive understanding of fluid mechanics problems, differential analysis of fluid flow is necessary. Therefore fluid mechanics governing equations and conservation of mass and momentum must be analyzed.

Conservation of mass (also called continuity equation) in vector notation is [1]:

$$\frac{\partial \rho}{\partial t} + \nabla \cdot (\rho V) = 0 \tag{4.4}$$

where ρ is the fluid density, V is the velocity vector, and ∇ is the gradient vector. The first and second terms in Eq. (4.2) denote the rate of change in fluid density and net rate of mass outflow per unit volume, respectively. Density for incompressible fluids is constant. Therefore Eq. (4.4) becomes:

$$\nabla \cdot V = 0 \tag{4.5}$$

For a carotid artery, blood vessel with one inlet and two outlets (see Fig. 4.8), the continuity Eq. (4.5) is simplified as [15]:

$$V_1 A_1 = V_2 A_2 + V_3 A_3 \tag{4.6}$$

or

$$Q_1 = Q_2 + Q_3 \tag{4.7}$$

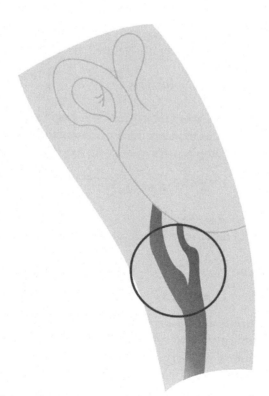

Figure 4.8 Carotid artery (schematic of a control volume with one inlet and two outlets).

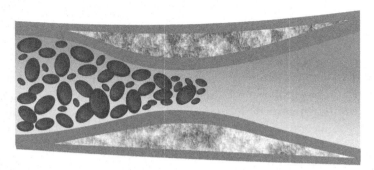

Figure 4.9 Schematic of a contraction or expansion in an artery.

where Q is the volume flow rate going into or out of the control volume. The continuity equation for expansion in artery (i.e., embolic as shown in Fig. 4.9) is:

$$V_1 A_1 = V_2 A_2 \qquad (4.8)$$

or

$$Q_1 = Q_2 = Q_3 \qquad (4.9)$$

Based on electrical analogy the resistance to blood flow, R_{flow}, through any vessels (i.e., blood flow in the circulatory system) is defined as:

$$R_{flow} = \frac{\Delta P}{Q} \qquad (4.10)$$

Therefore the pressure drops across a resistance vessel, a rigid vessel with constant volume $\Delta P = Q R_{flow}$. This implies that as resistance to flow increases the pressure drop across the vessel increases. Also, pressure drop across a vessel with length L and radius r is [16]:

$$\Delta P = R_{flow} Q = \frac{2\mu L Q}{\pi r} \qquad (4.11)$$

where the tube resistance is $R_{flow} = \frac{2\mu L}{\pi r}$.

Pressure drop across a compliance vessel, an elastic vessel with no noticeable resistance, is minimal.

($\Delta P = 0$) is minimal, so $P_{inlet} = P_{outlet} = P$ and volume is given by:

$$V = C_{flow}(P - P_{ext}) = C_{flow} P \qquad (4.12)$$

where P_{ext} is taken to be zero and C_{flow} is another important blood flow nondimensional parameter called the compliance. The compliance is change in volume caused by change in the pressure and is defined as [17]:

$$C_{flow} = \frac{\Delta V}{\Delta P} \qquad (4.13)$$

Another important nondimensional parameter that determines the change in volume caused by change in pressure is inertance, I_{flow} [17]:

$$I_{flow} = \frac{\Delta V}{\Delta Q} \qquad (4.14)$$

$$\Delta P = P_{inside} - P_{outside} = T\left(\frac{1}{r_1} + \frac{1}{r_2}\right) \qquad (4.15)$$

$$T = \sigma t$$

where r_1, r_2 is the vessel radius of curvature, σ is the circumferential stress, t is the thickness, and T is the circumferential tension (force per unit length along the tube, L) and has force/length units.

For a cylindrical vessel where $r_1 = r$ and $r_2 = \infty$, law of Laplace is as follows:

$$\Delta P = \frac{T}{r} \qquad (4.16)$$

And for a spherical vessel where $r_1 = r$ and $r_2 = r$, it reduces to:

$$\Delta P = \frac{2T}{r} \qquad (4.17)$$

Force per unit area distributed across the outer half of an inner curved arterial such as aorta wall (r_i is the curved inner radius) with cross-sectional area $\pi r_i L$ is $\rho u^2 A L / r_t$. For smaller artery with smaller effective area, $A = \pi r_i L/2$, the pressure is given by:

$$P = \frac{2\rho u^2 r_i}{r_t} \qquad (4.18)$$

4.1.3.1 Conservation of Momentum

Based on Newton's second law the conservation of momentum is [1]:

$$\nabla \cdot [-PI + \tau] + g = \rho\frac{\partial V}{\partial t} + \rho(V \cdot \nabla)V \qquad (4.19)$$

where P is the pressure, τ is the stress tensor, and I is the identity matrix. The first and the second terms on the left hand side of Eq. (4.17) denote the inertia force per unit volume and are local and convective acceleration, respectively. Based on Stokes theorem, shear stress for a Newtonian fluid is expressed as [2]:

$$\tau = 2\mu S - \frac{2}{3}\mu(V \cdot \nabla)I \qquad (4.20)$$

S is the strain-rate tensor and is [1]:

$$S = \frac{1}{2}\left(\nabla V + (\nabla V)^T\right) \qquad (4.21)$$

By plugging Eqs. (4.20) and (4.21) into the momentum equation (Eq. 4.19), it becomes [1]:

$$\nabla \cdot \left[-PI + \mu(\nabla V + (\nabla V)^T) - \frac{2}{3}\mu(V \cdot \nabla)I \right] + g = \rho \frac{\partial V}{\partial t} + \rho(V \cdot \nabla)V \quad (4.22)$$

where pressure (P) and velocity (V) are dependent variables. Eq. (4.22) is the general form of momentum equation and is used for both compressible and incompressible fluids. For incompressible fluids, such as gases, an additional equation, equation of state, is considered. The simplest form of equation of state is [1]:

$$\rho = \frac{pM}{R_u T} \quad (4.23)$$

where M is the molecular weight of the gas, R_u is the universal gas constant, and T is the absolute temperature.

For incompressible fluids ($\rho = $ constant) the third term from the left hand side of Eq. (4.23) vanishes and it reduces to [1]:

$$\nabla \cdot \left[-pI + \mu(\nabla V + (\nabla V)^T) \right] + g = \rho \frac{\partial V}{\partial t} + \rho(V \cdot \nabla)V \quad (4.24)$$

4.1.4 Euler and Bernoulli Equations

In the case of inviscid fluids ($\mu = 0$) the equation of motion (Eq. 4.22) is simplified to [18]:

$$\nabla P + g = \rho \frac{\partial V}{\partial t} + \rho(V \cdot \nabla)V \quad (4.25)$$

Eq. (4.25) is commonly referred to Euler's equation of motion. Under steady-state condition, the first term of right-hand side of the equation is to be zero and Euler's equation is simplified as follows [18]:

$$\nabla P + g = \rho(V \cdot \nabla)V \quad (4.26)$$

By integrating this equation along some arbitrary streamline and also assuming incompressible fluid we obtain [18]:

$$\frac{P}{\gamma} + \frac{V^2}{2g} + z = \text{const} \quad (4.27)$$

where γ is the specific weight of fluid and z is the elevation of the point above a reference plane. Eq. (4.27) is called Bernoulli equation and is valid for inviscid, steady, incompressible flow along a streamline or in the case of irrotational flow along any two arbitrary points. In other words, the Bernoulli equation indicates that the pressure stays constant during the flow when the tube cross-section and height do not change.

4.1.5 Reynolds Number

A fundamental nondimensional parameter in analyses of fluid flow is the Reynolds number [18]:

$$\mathrm{Re} = \frac{\rho U L}{\mu} \tag{4.28}$$

where U is the velocity scale and L denotes a representative length. The Reynolds number represents the ratio between inertial and viscous forces. At low Reynolds numbers, viscous forces dominate and tend to damp out all disturbances, which leads to laminar flow. At high Reynolds numbers, disturbances appear and at high enough Reynolds numbers the flow field eventually ends up in a chaotic state called turbulence.

Another important nondimensional number in biofluid mechanics is the Womersley number (α), relating the pulsatile flow frequency with viscous effects [19]:

$$\alpha = L\left(\frac{\omega}{\nu}\right)^{1/2} = (2\pi\,\mathrm{Re}\,\mathrm{St})^{1/2} \tag{4.29}$$

where St is the Strouhal number, a dimensionless number describing the mechanism of oscillating flow, and is given by [19]:

$$\mathrm{St} = \frac{fL}{U} \tag{4.30}$$

where f is the frequency of vortex shedding, and L and U are the characteristic length and the flow velocity.

4.1.6 Initial and Boundary Conditions

In general, the Navier–Stokes equation does not have exact solution due to the nonlinear term that exists in the equation of motion (convective acceleration). Typically a numerical scheme is used to analyze the Navier–Stokes equation. In some unique problems, like very-low-speed flow, the convective term drops out and the exact solutions become available [18]. To solve Navier–Stokes equation initial and boundary conditions must be available. The initial boundary condition is the condition of the system at time zero. Typical boundary conditions in fluid dynamic problems are: solid boundary conditions, inlet and outlet boundary conditions, and symmetry boundary conditions.

4.1.6.1 Solid Boundary Conditions

Solid boundary condition is also referred to as physical boundary condition or Dirichlet boundary condition. It is the most common boundary condition that describes the flow behavior at any solid boundary in any fluid dynamics problems. It is also called wall boundary condition. A typical wall boundary condition is no slip boundary condition in which the fluid adheres to the stationary wall and maintains zero velocity relative to the solid wall.

$$V = 0 \tag{4.31}$$

It hence implicitly assumes that there are no viscous effects at the wall and hence, no boundary layer develops. This condition is a reasonable approximation if the main effect of the wall is to prevent fluid from leaving the domain.

It is common to assume sliding wall boundary condition for a physical wall. Here the wall behaves like a conveyor belt; that is, the surface slides in its tangential direction. The wall does not have to actually move in the coordinate system [18]:

$$V = V_w t \tag{4.32}$$

where V_w is the velocity of the sliding wall and t is the unity tangential vector to the surface. If the wall moves, moving boundary condition applies and is formulated as [18]:

$$V = V_w \tag{4.33}$$

where V_w is the velocity vector of the moving wall.

4.1.6.2 Inlet and Outlet Boundary Conditions

To analyze any fluid dynamics problem, it is essential to know some of the flow characteristics such as velocity, mass flow rate, and pressure at the inlet and outlet of the medium. One of the most common inlet and outlet boundary conditions used are, respectively [18]:

$$V = -V_i n \tag{4.34a}$$

$$V = -V_o n \tag{4.34b}$$

where V_i and V_o are the inlet and outlet velocity, respectively. Minus sign appears because the flow direction is opposite to the velocity and unit normal vector direction at inlet and exit. In some cases the inlet mass flow rate is known [18]:

$$\dot{m}_{in} = -\int_A \rho(V \cdot n) dA \tag{4.35}$$

$\dot{m}_{in}$ is the rate of mass flow normal to the surface. Again, minus sign appears due to the flow direction, which is opposite to the velocity vector and unity normal vector to the surface at inlet. Another common inlet/exit boundary condition is the specified pressure:

$$P = P_0 \tag{4.36}$$

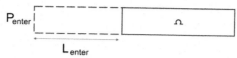

Figure 4.10 An example of the physical domain when using the laminar inflow boundary condition. Ω is the actual computational domain, whereas the dashed domain is an imaginary domain.

In many cases, the fluid is assumed to be inviscid at the inlet boundary, and momentum equation (Eq. 4.22) can be coupled with Eq. (4.34) to give [1]:

$$\left[\mu\left(\nabla V + (\nabla V)^T\right) - \frac{2}{3}\mu(V\cdot\nabla)I\right]n = 0 \tag{4.37}$$

For incompressible fluid Eq. (4.22) is simplified to [1]:

$$\left[\mu\left(\nabla V + (\nabla V)^T\right)\right]n = 0 \tag{4.38}$$

In most fluid dynamics problems the laminar inflow condition is used, as shown in Fig. 4.10. Assume that an imaginary domain of length L_{exit} is attached to the outlet of the computational domain. The domain is an extrusion of the outlet boundary, which means that laminar outflow requires the outlet to be flat. The boundary condition uses the assumption that the flow in this imaginary domain is fully developed laminar flow. The "wall" boundary conditions for the imaginary domain are inherited from the real domain, Ω [18].

Similarly, the laminar outflow condition corresponds to the situation shown in Fig. 4.11: assume that an imaginary domain of length L_{exit} is attached to the outlet of the computational domain. The domain is an extrusion of the outlet boundary, which means that laminar outflow requires the outlet to be flat. The boundary condition uses the assumption that the flow in this imaginary domain is fully developed laminar flow. The "wall" boundary conditions for the imaginary domain are inherited from the real domain, Ω [18].

4.1.6.3 Symmetric Boundary Conditions

Symmetric boundary condition, a combination of Dirichlet and Neumann boundary conditions, is the condition that no fluid penetrates across the boundary. Therefore flow across the boundary is zero:

$$V\cdot n = 0 \tag{4.39}$$

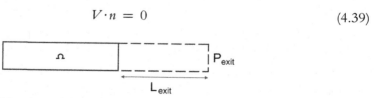

Figure 4.11 An example of the physical domain when using the laminar outflow boundary condition. Ω is the actual computational domain, whereas the dashed domain is an imaginary domain.

Using symmetric boundary condition reduces Eq. (4.22) to:

$$\left[-pI + \mu\left(\nabla V + (\nabla V)^T\right) - \frac{2}{3}\mu(V\cdot\nabla)I \right] n = 0 \tag{4.40}$$

4.2 NON-NEWTONIAN FLUID

Non-Newtonian fluids are fluids whose shear stress varies nonlinearly with rate of shear strain. In other words, the viscosity of non-Newtonian fluids is not constant and varies with the applied force. Some examples of non-Newtonian fluids are blood, heavy oil, starch suspensions, and shampoo. Blood (a suspension of red and white blood cells, platelets, and plasma) behaves as a non-Newtonian fluid in small arterial vessels (i.e., vessels with diameter less than 0.5 mm) where the shear rate is low (less than 20 s^{-1}) [20]. For large arterial vessels (vessels with diameter greater than 1 mm) where the shear rate is high, blood behaves like a Newtonian fluid [20]. Viscosity of some non-Newtonian fluids can also vary with time. They are sometimes referred to as memory materials. Some examples of such fluids are printer ink, paints, and gelatin. Detailed discussion of time varying viscosity is provided in the following section.

4.2.1 Classification of Non-Newtonian Fluid

In general, non-Newtonian fluids can be grouped into three general classifications: time-independent, time-dependent, and viscoelastic fluids.

4.2.1.1 Time-Independent Non-Newtonian Fluids

These are fluids in which the viscosity does not depend on fluid shear stress rate history. The time-independent non-Newtonian fluids are usually divided into shear thickening (sometimes referred to as dilatant), shear thinning (sometimes referred to as pseudoplastic), and typical Newtonian fluids. In shear thickening fluids the viscosity increases as stress increases [20]. An example of dilatant fluid is sand in water. On the other hand, when viscosity decreases when stress increases, the fluid is called pseudoplastic fluid. An example of pseudoplastic fluid is blood. This application is highly favored within the body, as it allows the viscosity of blood to decrease with increased shear strain rate [20]. Newtonian fluid is a special case of the non-Newtonian time-independent fluid where the viscosity is constant. An example of such fluid is blood plasma.

Some important time-independent non-Newtonian fluids are power law fluid, Bingham fluid, and some others (i.e., Herschel—Bulkley fluid, Casson fluid, Powell—Eyring fluid).

Bingham fluid, an important non-Newtonian fluid with linear shear stress/shear strain relationship, requires a finite yield stress before it begins to flow (the plot of shear stress against shear strain does not pass through the origin) [21]. Several examples are clay suspensions, drilling mud, toothpaste, mayonnaise, chocolate, and mustard. The surface of a Bingham plastic can hold peaks when it is still. By contrast, Newtonian fluids have flat featureless surfaces when still. The Bingham fluid constitutive equation is as follows [21]:

$$\tau = \mu\gamma' + \tau_0 \quad \tau \geq \tau_0$$
$$\gamma' = 0 \qquad \tau \leq \tau_0 \tag{4.41}$$

4.2.1.2 Time-Dependent Non-Newtonian Fluids

These are fluids that depend on strain rate as well as time of the applied shear. In other words, some non-Newtonian fluids show different viscosity values as they undergo shear stress. These fluids are divided into two groups: rheopectic, whose viscosity increases with time, and thixotropic, whose viscosity decreases with time. Rheopectic fluids are sometimes referred to as time-dependent dilatants and thixotropic fluids as time-dependent pseudoplastics. Some examples of rheopectic fluids are printer ink and gypsum paste and those of thixotropic fluids are paints and hydrogenated oil [21].

4.2.1.3 Viscoelastic Fluids

These are fluids that not only display viscous behavior but also are elastic, materials that obey Hooke's law. Some examples of such fluids are fluids with large polymeric molecules and whipped cream. The governing equation for viscoelastic fluids is:

$$\gamma = \frac{\tau}{\mu_0} + \frac{\tau}{\lambda} \tag{4.42}$$

where (λ) is a rigidity modulus. Table 4.4 summarizes the types of non-Newtonian fluids [22].

4.2.2 Non-Newtonian Fluid Flow in Human Body

Body fluids, sometimes referred to as biofluids, are liquids inside the human body and account for about 70% of human body weight. Water is the dominant portion of the body fluids and its weight varies depending on the age, gender, body fat, and some other parameters [21].

Intracellular fluid is the portion of body fluid that is located within the human cells. It contains about 65% of the biofluid. The rest of the body fluid is located outside of the human cells and is referred to as extracellular fluid (ECF). The ECF components are: interstitial fluid (fluid between the cells), lymph (fluid that circulates throughout the lymphatic system), and blood.

4.2.2.1 Interstitial Fluid

Also referred to as tissue space, interstitial fluid is the fluid that is placed between the cells and provides the body with nutrients. It accounts for 80% of the body fluid, about 10 L

Table 4.4 Summary of the types of non-Newtonian fluids [22]

Group	Type	Definition	Example
Time-independent viscosity	Shear thickening (dilatant)	Fluid is one in which viscosity increases with the rate of shear strain. This behavior is only one type of deviation from Newton's law, and it is controlled by factors such as particle size, shape, and distribution. The properties of these suspensions depend on Hamaker theory and van der Waals forces and can be stabilized electrostatically or sterically	*Corn starch*: when stirred slowly it looks milky, when stirred vigorously it feels like a very viscous liquid
	Shear thinning (pseudoplastic)	Shear thinning is an effect where a fluid's viscosity decreases with an increasing rate of shear strain	Blood as lava, ketchup, cream, blood, paint; it is also a common property of polymer solutions and molten polymers
Time-dependent viscosity	Rheopecty	Rheopecty or rheopexy is the rare property of some non-Newtonian fluids to show a time-dependent increase in viscosity (time-dependent viscosity); the longer the fluid undergoes shearing force, the higher its viscosity. Rheopectic fluids, such as some lubricants, thicken or solidify when shaken	Printer ink, gypsum paste
	Thixotropy	Fluids show a time-dependent change in viscosity; the longer the fluid undergoes shear stress, the lower its viscosity	Yogurt, gelatin gels, hydrogenated castor oil, some drilling muds, many paints, many colloidal suspensions
Viscoelastic	Kelvin material, Maxwell material	"Parallel" linearistic combination of elastic and viscous effects [1]	Some lubricants, whipped cream, Silly Putty

(2.4 imperial gallons or ~2.9 US gal). It is very similar to plasma. Interstitial fluid also contains transcellular fluid [23].

4.2.2.2 Intravascular Fluid

This is also referred to as blood plasma, is the amount of fluid inside the human blood vessel, and serves as the protein reserve of the human body. It accounts for 55% of the body's total blood volume and about 20% of ECF. Plasma is an aqueous solution of proteins, organic molecules, and minerals in which the blood cells are suspended. Blood plasma is 95% water by volume and contains dissolved proteins (i.e., albumins, globulins, and fibrinogen), glucose, clotting factors, electrolytes (Na^+, Ca^{2+}, Mg^{2+}, HCO^{3-}, Cl^-, etc.), hormones, and carbon dioxide [23]. In general blood contains 55% plasma and 45% blood cells, the red and white blood cells and the platelets [24]. The density of Blood plasma is approximately 1025 kg/m^3, or 1.025 g/mL [24].

4.2.2.3 Lymphatic Fluid

This is the fluid that enters the lymph vessels by infiltration. Lymph is the fluid that circulates throughout the lymphatic system and is formed when the interstitial fluid is collected through lymph capillaries. The lymph ultimately mixes back with blood by traveling into the right or the left subclavian. The lymph composition constantly changes because the blood and the surrounding cells continually exchange substances with the interstitial fluid. There is approximately 6—10 L of lymph in the body, compared with 3.5—5 L of blood [24].

4.2.2.4 Transcellular Fluid

This is the portion of the total body fluid that is not inside the cells (about 3% of total body water). It is contained within epithelial-lined spaces and is separated from plasma and interstitial fluid by cellular barriers. The composition and function of transcellular fluid vary inside the human body. For instance, it serves as lubrication in joints, whereas the removal of electrolytes and molecules from the body is done by the urine. Examples of this fluid are cerebrospinal fluid, and ocular fluid, joint fluid, and bladder urine [2].

4.2.3 Non-Newtonian Fluid Flow Models for Blood

Blood is a heterogeneous multiphase mixture of solid corpuscles (red blood cells, white blood cells, and platelets) suspended in plasma. The composition of blood, as well as the properties of the constituents, causes the blood to have a complex macroscopic behavior. Blood can be modeled as Newtonian and non-Newtonian fluid depending on its application.

There exist no general shear rate values that determine the transition from Newtonian to non-Newtonian fluid. It depends on the size and shape of the vessel as well as the nature of the blood transportation processes in different parts of the human body. In general, if the shear rate is larger than 100 s^{-1}, blood is assumed to behave as Newtonian fluid. For example, in large cavities such as the ventricles and atria inside the myocardium as well as the large arteries and veins, the blood essentially behaves as a Newtonian fluid [2].

For shear rates less that 100 s^{-1} one must assume that the blood behaves as non-Newtonian fluid. Therefore any reliable blood flow models used here must take into account its non-Newtonian characteristics. Typical examples of non-Newtonian effect are but not limited to: large blood vessels that mainly apply to arteries and veins, small blood vessels that broadly include capillaries and possibly arterioles, and porous tissue such as the myocardium and muscles in general; refer to Fig. 4.12.

Several mathematical and computational models have been used to describe the flow of blood in vessels. The most referred ones in Newtonian fluid are the elastic one-dimensional Navier—Stokes and the rigid Hagen—Poiseuille flow. The most important generalized non-Newtonian models are: Carreau—Yasuda, Casson, power law, Cross, Herschel—Bulkley, Oldroyd-B, Quemada, Yeleswarapu, Bingham, Eyring—Powell, and Ree—Eyring [3]; see Table 4.5.

4.3 BLOOD FLOW IN ARTERIES AND VEINS

The circulatory system, also called the cardiovascular system, is made up of heart, blood vessels, and blood. Blood vessels are divided into veins and arteries. Veins carry blood toward the heart for purification and arteries carry oxygenated blood from the heart to various parts of the body. Veins are about 1 mm to 1—1.5 cm in diameter and capillaries

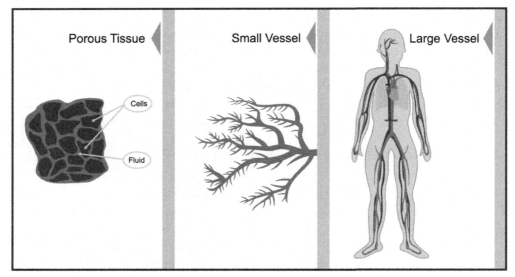

Figure 4.12 The three subsystems of blood vessels.

Table 4.5 Constitutive equations of the rheological models [20]

	Rheological model	Effective viscosity
Casson type	Casson	$\eta(\dot\gamma) = \left(\sqrt{\eta_c} + \sqrt{\tau_c/\dot\gamma}\right)^2$, $\quad \eta_c = 0/00414$, $\quad \tau_c = 0/0038$
	K-L	$\eta(\dot\gamma) = \frac{1}{\dot\gamma}\left[\tau_c + \eta_c\left(\alpha_2\sqrt{\dot\gamma} + \alpha_1\dot\gamma\right)\right]$, $\quad \eta_c = 0/0035$, $\tau_c = 0/005$, $\alpha_1 = 1$, $\alpha_2 = 1/19523$
	Modified Casson	$\eta(\dot\gamma) = \left(\sqrt{\eta_c} + \frac{\sqrt{\tau_c}}{\sqrt{\lambda} + \sqrt{\dot\gamma}}\right)^2$, $\quad \eta_c = 0/002982$, $\quad \tau_c = 0/02876$, $\quad \lambda = 4/020$
Carreau type	Carreau	$\eta(\dot\gamma) = \eta_\infty + (\eta_0 - \eta_\infty)\left[1 + (\lambda\dot\gamma)^2\right]^{(n-1)/2}$, $\quad \eta_\infty = 0/0035\text{Pas}$, $\eta_0 = 0/056\text{Pas}$, $\lambda = 3/313005$, $n = 0/3568$
	Carreau–Yasuda	$\eta(\dot\gamma) = \eta_\infty + (\eta_0 - \eta_\infty)\left[1 + (\lambda\dot\gamma)^a\right]^{(n-1)/a}$, $\quad \eta_\infty = 0/0035\text{Pas}$, $\eta_0 = 0/16\text{Pas}$, $\lambda = 8/2$, $n = 0/2128$, $a = 0/64$
	Cross	$\eta(\dot\gamma) = \eta_\infty + \frac{\eta_0-\eta_\infty}{1+(\lambda\dot\gamma)^n}$, $\quad \eta_\infty = 0/0035\text{Pas}$, $\eta_0 = 0/0364\text{Pas}$, $\lambda = 0/38$, $a = 1/45$
Power law type	Power law	$\eta(\dot\gamma) = k(\dot\gamma)^{n-1}$, $\quad k = 0/017$, $n = 0/708$
	Modified power law	$\eta_\infty < \eta(\dot\gamma) = k(\dot\gamma)^{n-1} < \eta_0$, $\quad k = 0/017$, $n = 0/708$, $\eta_\infty = 0/0035\text{Pas}$, $\eta_0 = 0/056\text{Pas}$
	Generalized power law	$\eta(\dot\gamma) = k(\dot\gamma)\dot\gamma^{n(\dot\gamma)-1}$, $\quad k(\dot\gamma) = \eta_\infty + \Delta\eta\exp\left[-\left(1 + \frac{\dot\gamma}{a}\right)\exp\left(\frac{-b}{\dot\gamma}\right)\right]$, $n(\dot\gamma) = n_\infty + \Delta n\,\exp f\left[-\left(1 + \frac{\dot\gamma}{c}\right)\exp\left(\frac{-d}{\dot\gamma}\right)\right]$, $\eta_\infty = 0/0035\text{Pas}$, $\Delta\eta = 0/025$, $a = 50$, $b = 3$, $n_\infty = 1/0$, $\Delta n = 0/45$, $c = 50$, $d = 4$

are about 5—10 μm in diameter. The size of arteries varies depending on their types: muscular and elastic. The diameter of elastic arteries is greater than 1 cm and that of muscular arteries is about 0.1—10 mm. Systemic arteries deliver blood to the arterioles, and then to the capillaries, where nutrients and gasses are exchanged.

The blood flowing through the veins and elastic arteries (tube larger than 1 mm) behaves as Newtonian fluid. For blood flowing through tubes less than approximately 1 mm in diameter (i.e., capillary and muscular arteries), the viscosity is not constant and therefore blood behaves as a non-Newtonian fluid [23].

In the following subsections the mathematical models for different parts of the cardio-vascular system are described.

4.3.1 Arteries

Arteries are the high-pressure blood vessels that carry blood away from the heart, through increasingly smaller arteries, to arterioles, and further to the level of capillaries. Blood flow in the systemic arteries is usually assumed to be one-dimensional and is modeled from the Navier—Stokes equations as [25]:

$$\frac{\partial Q}{\partial t} + \frac{\partial}{\partial x}\left(\beta \frac{Q^2}{A}\right) = -\frac{A}{\rho}\frac{\partial P}{\partial x} - \frac{\pi D}{\rho}\tau_0$$

$$\frac{\partial A}{\partial t} + \frac{\partial Q}{\partial x} = 0$$

(4.43)

where Q is the volumetric flow rate ($Q = UA$), β is the momentum correction factor, P is the mean pressure, ρ is the blood density, A is the cross-sectional area of the artery, and D is the diameter. The viscous shear stress acting on the arterial wall, τ_0, is given by [25]:

$$\tau_0 = f_{darcy}\frac{\rho \breve{u}\left|\breve{u}\right|}{8}$$

(4.44)

where f_{darcy} is the Darcy friction factor and $\breve{u}$ is the mean value of the axial velocity.

Pressure decreases as blood moves from the arteries to veins. Blood pressure is highest in the arteries and lowest in the veins. Typical normal values of blood pressure, systolic pressure (the higher value) over diastolic pressure (the lower value), for a 70-kg man is 120/80 [22]. Pulse pressure, $P_{pulsatile}$, the change in amplitude of the pulse pressure wave (the difference between systole and diastole pressure), is calculated as [26]:

$$P_{pulsatile} = P_{systolic} - P_{diastolic}$$

(4.45)

The mean arterial blood pressure (P_{mabp}) assuming that the systole lasts for about one-third of the cycle and diastole for about two-thirds is:

$$P_{mabp} = \frac{1}{3}\left(P_{systole} + 2P_{diastole}\right)$$

(4.46)

The relation between blood pressure and blood flow in the aorta or any pulmonary artery is given by Windkessel (2, 3, and 4 element) models. Models assume that cardiac cycle starts at systole; the period of the systole is 0.4 of the period of cardiac cycle. It also assumes that Arterial Compliance, Peripheral Resistance, and Inertia are modeled as a capacitor, a resistor, and an inductor, respectively. The 2-element diagram (see Fig. 4.13) and equation Windkessel theoretical models are given as, respectively [2]:

$$I(t) = \frac{P(t)}{R} + C\frac{dP(t)}{dt} \tag{4.47}$$

The analytical solution to the 2-element Windkessel model (Eq. 4.47) during systole (nonhomogenous solution) gives:

$$y(t) = c_1 \exp\left(\frac{-t}{RC}\right) + \frac{-\exp\left(\frac{t}{RC}\right)T_s I_0 R\left(C\pi R\cos\left(\frac{\pi t}{T_s}\right) - T_s \sin\left(\frac{\pi t}{T_s}\right)\right)}{T_s^2 + C^2\pi^2 R^2} \tag{4.48}$$

The homogenous solution (diastole) of Eq. (4.47) gives:

$$P(t) = c\exp\left(\frac{-t}{RC}\right) \tag{4.49}$$

Constants (c_1 and c) are determined by appropriate boundary conditions.

4.3.1.1 Blood Flow Through Aorta

There is no blood flow in the aorta during the diastole, and during the systole the blood flow into the aorta ($I(t)$) is usually modeled as sinusoidal wave [27]:

$$I(t) = I_0 \sin\left(\frac{\pi \cdot Rm(t, T_c)}{T_s}\right) \tag{4.50}$$

where I_0 is the amplitude of blood flow during systole, T_c and T_s are the period of the cardiac cycle and systole in seconds, respectively, t is time in seconds, and $Rm(t, T_c)$ represents the remainder of t/T_c. The maximum amplitude of blood flow during

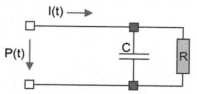

Figure 4.13 The 2-element diagram of Windkessel theoretical models.

systole, where $T_c = 2.5\ T_s$ and the blood flow in one cardiac cycle is 90 cm^3, is 424.1 mL ($I_{0,max} = 424.1$ mL).

4.3.2 Veins

The blood in veins is modeled as Newtonian fluid and the flow is assumed to be incompressible, homogenous, and laminar in a tube. The mass and momentum equations for veins are, respectively [27]:

$$\frac{\partial R}{\partial t} + V\frac{\partial R}{\partial x} + \frac{R}{2}\frac{\partial V}{\partial x} = 0$$

$$\frac{\partial V}{\partial t} + (2\alpha - 1)V\frac{\partial V}{\partial x} + 2(\alpha - 1)\frac{V^2}{R}\frac{\partial R}{\partial x} + \frac{1}{\rho}\frac{\partial P}{\partial x} + \left(\frac{2\nu\alpha}{\alpha - 1}\right)\frac{V}{R^2} = 0$$

(4.51)

where V is the mean velocity of the profile, R is the vessel radius, P is the fluid pressure, and $\rho\ \nu$ are fluid density and kinematic viscosity, respectively, and α is the velocity profile parameter. Velocity profile is given by:

$$u = \left(\frac{\gamma + 2}{\gamma}\right)V\left[1 - \left(\frac{r}{R}\right)^\gamma\right]$$

(4.52)

where $\gamma = \frac{2-\alpha}{\alpha-1}$.

4.3.3 Capillaries

The capillary diameter is about 8 μm on average and vessel wall is about 1 μm in thickness (a tube of a single layer of rolled-up endothelial cells). The wall surface area is 300–1200 μm^2. The entire capillary wall surface (about 313 m^2) comprises up to 0.25–1 trillion endothelial cells. In general the blood behaves as non-Newtonian fluid in capillaries.

For human red blood cells to pass through any vessel, the vessel must be at least 2.7 μm in diameter. The minimum tube diameter $D_{t,min}$ is given by [28]:

$$V_{MC} = \left(\frac{A_{MC}}{4}\right)D_{t,\min} - \left(\frac{\pi}{12}\right)D_{t,\min}^3$$

(4.53)

where A_{MC} is the mean cell surface area and V_{MC} is the mean cell volume.

Blood flow Reynolds number in capillary tubes is much less than 1 and therefore the Poiseuille's law is applied. The basic flow equation that applies to a single vessel as well as a network of vessels (i.e., the vascular bed of an organ or even the entire systemic circulatory system) is:

$$Q = \frac{\Delta P}{R}$$

(4.54)

where Q is the flow rate (volume/time), ΔP is the pressure difference (mm Hg), and R is the resistance to the flow (mm Hg × time/volume) and is as follow:

$$R = \frac{8l\mu}{\pi r^4} \tag{4.55}$$

where r is the inner radius of the vessel, L is the vessel length, and μ is the blood viscosity. Combining Eqs. (4.54) and (4.55) results in the well-known Poiseuille equation:

$$Q = \frac{\pi r^4 \Delta P}{8L\mu} \tag{4.56}$$

The transvascular flow rate is determined by Starling's law as follows [29]:

$$Q_{trans} = \pi DLL_P \left[P_b - P_i - \sigma_s \left(P_{opp} - P_{opi} \right) \right] \tag{4.57}$$

where P_i is the interstitial pressure, P_b is the intravascular pressure, P_{opp} is the osmotic pressure of the plasma, P_{opi} is the osmotic pressure of the interstitial fluid, L_P is the hydraulic conductivity of the vessel wall, and σ_s is the average osmotic reflection coefficient for the plasma proteins. It is noted that Starling's law represents the role of hydrostatic and oncotic pressures in the movement of fluid across capillary membranes. The apparent blood viscosity is given by [3]:

$$\mu_{app} = \mu_{plasma} \cdot \mu_{rel} \tag{4.58}$$

The relative apparent viscosity, μ_{rel}, as a function of the tube diameter and hematocrit is given by Pries et al. and is as follows [30]:

$$\mu_{rel} = \left[1 + (\mu_{45} - 1) \left(\frac{(1-H)^C}{(1-0.45)^C - 1} \right) \left(\frac{D}{D-1.1} \right)^2 \right] \left(\frac{D}{D-1.1} \right)^2 \tag{4.59}$$

where D is the vessel diameter (in μm), μ_{45} is the relative apparent blood viscosity for a fixed hematocrit of 0.45, C describes the shape of viscosity dependency on the hematocrit and are given by, respectively [27]:

$$\mu_{45} = 6 \exp(-0.085D) + 3.2 - 2.44 \exp\left(-0.06D^{0.645}\right) \tag{4.60}$$

$$C = (0.8 + \exp(-0.075D)) \left(-1 + \frac{1}{1 + 10^{-11}D^{12}} \right) + \frac{1}{1 + 10^{-11}D^{12}} \tag{4.61}$$

The hematocrit and vessel diameter are dependent on blood flow characteristics such as velocity, wall shear stress, and pressure in vessels.

4.4 PULSATILE FLOW

The blood is pumped out by the heart has a highly pulsatile flow with periodic variations. Pulsatile blood flow transmits more energy to the microcirculation, which reduces the critical capillary closing pressure, augments lymph flow, reduces vasoconstriction reflexes and neuroendocrine responses, and improves tissue perfusion and cellular metabolism.

The pulsating characteristics are the result of two pumping mechanisms. First, the heart pump causes the blood to flow and velocity to oscillate from zero to very high rates. It happens as the valves at the entrances and exits to the ventricles intermittently close and open with each beat of the heart. Second, pumping is a result of the respiratory and skeletal systems, which exert their greatest action on venous flow []. Blood releases from the left ventricle are highly pulsatile, and pressure and blood flow are nonlinear and transient. This creates complex pulse patterns throughout the rest of the network and varies the shear stress that is applied to the layer of endothelial cells covering the vessel wall. Depending upon the amount of stress, the endothelial cells release chemicals that induce either dilation or constriction of the smooth muscle surrounding the vessel.

It is nearly impossible to mathematically model such a flow using the standard Navier–Stokes equations. Therefore the Womersley number is used for pulsatile flow. As discussed earlier, this dimensionless number has been developed to give a measure of the frequency and magnitude of pulsations rather than a model of the actual flow. The Wormesley number is primarily influenced by the size of the vessel since the blood density and viscosity remain fairly constant (with slight variations throughout).

The pressure that supplies blood to the body is broken into systolic and diastolic pressures. Systolic blood pressure is the pressure in the vascular system when the ventricles of the heart are contracting. The diastolic pressure is the pressure in the vascular system when the ventricles are relaxing. This constant pressure prevents the vessels from collapsing or experiencing blood backflow. Normal values for systolic and diastolic pressures are approximately120 and 80 mm Hg, respectively [2].

The continuity and Navier–Stokes equation for pulsatile flow are simplified as, respectively:

$$\frac{\partial u}{\partial x} = 0 \tag{4.62}$$

$$\rho \frac{\partial u}{\partial t} = -\frac{\partial P}{\partial x} + \mu \left(\frac{\partial^2 u}{\partial r^2} + \frac{1}{r} \frac{\partial u}{\partial r} \right) \tag{4.63}$$

The pressure gradient and the velocity profile for a pulsatile flow is given by, respectively:

$$\frac{\partial P}{\partial x} = \sum_{n=-N}^{N} C_n \exp(inwt) \tag{4.64a}$$

$$u(r,t) = \sum_{n=-N}^{N} U_n \exp(inwt) \tag{4.64b}$$

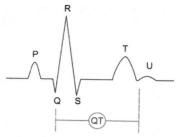

Figure 4.14 Electrocardiogram signal of a heart beat.

Substituting Eqs. (4.64a) and (4.64b) into the Navier–Stokes equation gives:

$$i\rho n w U_n = -C_n + \mu\left(\frac{\partial^2 u}{\partial r^2} + \frac{1}{r}\frac{\partial u}{\partial r}\right) \tag{4.65}$$

where the general solution is [2]:

$$U_n(r) = A_n J_0\left(\alpha\frac{r}{R}n^{1/2}i^{3/2}\right) + B_n Y_0\left(\alpha\frac{r}{R}n^{1/2}i^{3/2}\right) + \frac{iC_n}{\rho n w} \tag{4.66}$$

where k is a constant, A_n and B_n are arbitrary constants, $J_0(kr)$ is the Bessel function of first kind and order zero, $Y_0(kr)$ is the Bessel function of second kind and order zero, and α is the dimensionless Womersley number.

4.4.1 Characteristics of Pulsatile Flow

The heart produces a signature type of flow through the vascular system. The pulse is created by the de- and repolarization of the atria and ventricles. The electrical signal of a heart's pulsatile flow as detected by an electrocardiogram is shown in Fig. 4.14 [27]. The *P* wave is the depolarization of the atria, the QRS wave is the depolarization of the ventricles, and the *T* wave represents the repolarization of the ventricles.

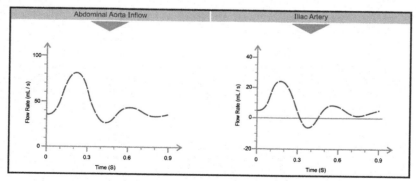

Figure 4.15 Flow rate curves for abdominal aorta and iliac arteries.

Fig. 4.15 depicts the flow pattern, repeating curve, for the abdominal aorta and iliac arteries [27]. As shown the maximum point on the curve is the systolic pressure when the ventricles of the heart contract. However, the minimum point on the curve means the diastolic pressure is retained in the arteries.

REFERENCES

[1] R.W. Fox, A.T. McDonald, Introduction to Fluid Mechanics, fourth ed., John Wiley & Sons, Inc.
[2] L. Waite, J. Fine, Applied Biofluid Mechanics, Mc-Graw Hill, 2007, http://dx.doi.org/10.1036/0071472177.
[3] A. Shahidian, Flow and Thermal Analysis of Bio-nano Fluid in a Channel like Vessel under the Effect of Electromagnetic Forces (MHD), Submitted in partial fulfillment of the requirements for the Ph.D. degree in Mechanical Engineering, K.N. Toosi University of Technology, January 2011.
[4] B. Alberts, Table 22−1 blood cells, Molecular Biology of the Cell (2012). NCBI Bookshelf (Chapter 22: Blood Vessels and Endothelial Cells).
[5] A. Waugh, A. Grant, Anatomy and Physiology in Health and Illness, tenth ed., Churchill Livingstone Elsevier, 2007, ISBN 978-0-443-10102-1, p. 22 (Chapter 2).
[6] J.N. Mazumdar, Biofluid Mechanics, World Scientific Publishing, 1992.
[7] The Franklin Institute Inc., Blood − The Human Heart.
[8] Medical Encyclopedia: RBC Count, Medline Plus.
[9] A. Guyton, J. Hall, Guyton Textbook of Medical Physiology, tenth ed., 2000, ISBN 072168677X.
[10] Deleted in review.
[11] L. Waite, Biofluid Mechanics in Cardiovascular Systems, Mc-Graw Hill, 2006.
[12] K. Saladin, Anatomy Physiology: The Unity of Form and Function, McGraw Hill, 2012.
[13] Histology image:22401lba from Vaughan, Deborah, A Learning System in Histology: CD-ROM and Guide, Oxford University Press, 2002, ISBN 978-0195151732.
[14] M. Pavelka, J. Roth, Functional Ultrastructure: An Atlas of Tissue Biology and Pathology, Springer, 2005, p. 232.
[15] Y.A. Cengel, J.M. Cimbala, Fluid Mechanics, Fundamental and Applications, third ed., McGraw Hill, 2010.
[16] J.R. Cameron, Medical Physics, John Wiley, 1978.
[17] M. Catanho, M. Sinha, V. Vijayan, Model of Aortic Blood Flow Using the Windkessel Effect, BENG 221-Mathematical Methods in Bioengineering Report, October 25, 2012.
[18] G. Falkovich, Fluid Mechanics, Cambridge University Press, 2011.
[19] J.R. Womersley, Method for the calculation of velocity, rate of flow and viscous drag in arteries when the pressure gradient is known (PDF), Journal of Physiology 127 (3) (March 1955) 553−563. PMC 1365740. PMID: 14368548.
[20] C. Tropea, A.L. Yarin, J.F. Foss, Springer Handbook of Experimental Fluid Mechanics, Springer, 2007, ISBN 978-3-540-25141-5, pp. 661−676.
[21] L.L. Schramm, Emulsions, Foams, and Suspensions: Fundamentals and Applications, Wiley VCH, 2005, ISBN 978-3-527-30743-2, p. 173.
[22] M.A. Rao, Rheology of Fluid and Semisolid Foods: Principles and Applications, second ed., Springer, 2007, ISBN 978-0-387-70929-1, p. 8.
[23] Tuskegee University (May 29, 2013) Blood, in: Dorland's Illustrated Medical Dictionary, thirty second ed., Elsevier, 2012, ISBN 978-1-4160-6257-8, p. 951 (Chapter 9), tuskegee.edu.
[24] The Physics Factbook − Density of Blood.
[25] G. Pedrizzetti, K. Perktold, Cardiovascular Fluid Mechanics, Springer-Verlag, 2003.
[26] R.E. Klabunde, Cardiovascular Physiology Concepts: Mean Arterial Pressure.
[27] D. Elad, S. Einav, Physical and flow properties of blood, in: Standard Handbook of Biomedical Engineering and Design, Mc-Graw Hill, 2004 (Chapter 3).

[28]   S.L. Schrier, S.A. Landaw, Mean Corpuscular Volume, September 30, 2011. uptodate.com.
[29]   J. West, Respiratory Physiology: The Essentials, ninth ed., Lippincott Williams & Wilkins, Baltimore, 2012, ISBN 978-1-60913-640-6, p. 177.
[30]   T.W. Secomb, A.R. Pries, Blood viscosity in microvessels: experiment and theory, Comptes Rendus Physique 14 (6) (2013) 470—478, http://dx.doi.org/10.1016/j.crhy.2013.04.002.

CHAPTER 5

Bio-Nanofluid Simulation

5.1 NANOFLUID

Fluids with suspended nanoparticles (below 100 nm) are called nanofluids, a term first proposed by Choi of the Argonne National Laboratory, USA, in 1995 [1]. The nanoparticles used in nanofluids are typically made of metals, oxides, carbides, or carbon nanotubes. Common base fluids include water, ethylene glycol, and oil. Nanofluids are considered to be the next-generation heat transfer fluids as they offer exciting new possibilities to enhance heat transfer performance compared with pure liquids. Recent documents indicate that nanofluids (such as water or ethylene glycol) with CuO or Al_2O_3 nanoparticles exhibit enhanced thermal conductivity [1]. Thus the use of nanofluids results in energy savings and facilitates the trend of device miniaturization. More exotic applications of nanofluids can be envisioned in biomedical engineering and medicine in terms of optimal nanodrug targeting and implantable nanotherapeutics devices [2]. Therefore the knowledge of bio-nanofluid thermophysical properties (such as blood with nanoparticles) becomes essential when flow and heat transfer study of blood in drug delivery and cancer therapy is considered.

The enhancement of thermal conductivity achieved in nanofluids is much greater than what has been predicted by conventional theories such as Maxwell [3] or Hamilton and Crosser [4]. Several experimental studies have explained the reason behind the enhancement of effective thermal conductivity such as the effect of the solid/liquid interfacial layer and the Brownian motion [2,5–9].

Four possible causes of increase in thermal conductivity given by Keblinski et al. [10] are: the Brownian motion of the nanoparticles, the molecular-level layering of the liquid at the liquid/particle interface, the nature of heat transport in the nanoparticles, and the effects of nanoparticle clustering. In another study, Keblinski et al. [11] simply reviewed and discussed the properties of nanofluids and future challenges.

5.1.1 Production of Nanofluid

Nanofluids are fluids with different nanoparticles (particles smaller than 100 nm) and base fluids. Many types of nanoparticles such as metals (Cu, Ag, Au), oxide ceramics (Al_2O_3, CuO), carbon nanotubes, and carbide ceramics (SiC, TiC) and various liquids such as water, oil, and ethylene glycol are used. Two general methods are used to produce nanofluids: two-step and one-step methods. In the two-step method the nanoparticles are made separately and dispersed in base fluid by different method. Ultrasound and high-shear dispersion techniques are used to produce nanofluids with oxide nanoparticles by

Nano and Bio Heat Transfer and Fluid Flow
ISBN 978-0-12-803779-9, http://dx.doi.org/10.1016/B978-0-12-803779-9.00005-4

the two-step method. Nanoparticles are made and dispersed in a fluid in a single process in the one-step method. This method is used to produce nanofluids in small quantities for research purposes. For nanofluids with high-conductivity metal nanoparticles, the one-step method is preferred.

The study of nanoparticles is done based on one of the following methods:

1. Transmission electron microscopy: This method provides the particle size distribution (mean diameter and standard deviation), the crystallinity of a sample, and the information about particle shape and phase transition.
2. Optical spectroscopies: This is a method suitable for metal nanoparticles.
3. X-ray diffraction: This method is used to understand the properties of synthesized materials.
4. Infrared spectroscopy: This is a good method to investigate the molecular vibrations in a nanomaterial. It is used to study the monolayers on the surface of nanoparticles.
5. Zeta potential: This method is used to understand the stability of colloidal and nanoparticle solutions. The minimum zeta potential for stability of a colloidal and nanoparticle solution is ±30 mV. The zeta potential is measured by microelectrophoresis method.

5.1.2 Nanofluid Properties

Nanofluid properties depend on various parameters such as base fluid properties, particle dimension and geometry, particle distribution, and fluid—particle interfacial effects. The detail of the nanofluid microstructures is necessary to estimate their effective properties. Two approaches are used to estimate the nanofluid properties when their detailed data are not available: determine the upper and the lower bounds on the effective properties from partial statistical information or estimate properties based on reasonable assumptions on the fluid microstructure.

Nanofluid density, ρ_{eff}, is calculated based on mixture rule as follows [12]:

$$\rho_{eff} = (1 - \varphi)\,\rho_f + \varphi\,\rho_p \tag{5.1}$$

In addition to mixture rule the mass averaging methods are used to calculate the specific heat of nanofluid, $C_{p,eff}$, as follows:

$$C_{p,eff} = \frac{(1 - \varphi)(\rho C_p)_f + \varphi(\rho C_p)_p}{(1 - \varphi)\rho_f + \varphi\,\rho_p} \tag{5.2}$$

Nanofluid viscosity, μ_{eff}, is calculated by the Einstein relation for nanoparticle volume concentration less than 0.03 as:

$$\mu_{eff} = \mu_f(1 + 2.5\,\varphi) \tag{5.3}$$

where ρ, C_p, μ, k, and φ are density, specific heat capacity, viscosity, thermal conductivity, and volume fraction, respectively. Also the subscripts f, p, and eff

represent the fluid, nanoparticles, and effective of nanofluid, respectively. Other important relations for nanofluid viscosity are listed below [12]:

$$\mu_{eff} = \mu_f \left[\frac{1}{1.25\varphi + 1.552\ \varphi^2} \right] = \mu_f \{ 1 + 2.5\ \varphi + 4.698\ \varphi^2 + ... \} \qquad (5.4)$$

$$\mu_{eff} = \mu_f \{ 1 + 2.5\ \varphi + 7.349\ \varphi^2 + ... \} \qquad (5.5)$$

$$\mu_{eff} = \mu_f \{ 1 + 2.5\ \varphi + 6.2\ \varphi^2 \} \qquad (5.6)$$

$$\mu_{eff} = \mu_f \{ 13.47 \exp (35.98\ \varphi) \} \qquad (5.7)$$

$$\mu_{eff} = \mu_f \{ 1 + 7.3\ \varphi + 123\ \varphi^2 \} \qquad (5.8)$$

Four different approaches are used to estimate the effective thermal conductivity, one of the most important thermophysical properties of nanofluid, as follows:

1. The mixture rule: The general mixture rule [13], parallel mixture rule, and series mixture rule [14] are written as follows, respectively:

$$k_{eff}^n = (1 - \varphi) k_f^n + \varphi\ k_p^n \quad 1 \leq n \leq 1 \qquad (5.9)$$

$$k_{eff} = (1 - \varphi) k_f + \varphi k_p \quad n = 1 \qquad (5.10)$$

$$k_{eff} = \frac{1}{(1 - \varphi) k_f^{-1} + \varphi\ k_p^{-1}} \quad n = -1 \qquad (5.11)$$

2. Maxwell's equation (for law particle volume concentration)

$$k_{eff} = k_f + 3\varphi \frac{k_p - k_f}{2k_f + k_p} \qquad (5.12)$$

3. Interfacial thermal resistance: is a special particle shell with zero volume concentration. It is calculated by the Benveniste [15] equation for spherical particle of radius r_p, which is embedded in base fluid with interfacial thermal resistance R as follows:

$$k_{eff} = k_f + 3\varphi \frac{k_p^R - k_f}{2k_f + k_p^R - \nu\varphi_p \left(k_p^R - k_f \right)} k_f \qquad (5.13)$$

where k_{eff}, k_p, k_f, and φ are the nanofluid effective thermal conductivity, nanoparticle thermal conductivity, base fluid thermal conductivity, and nanoparticle volume concentration, respectively.

4. Kumar [16] presented a dynamic model (moving particle model) for nanofluid effective thermal conductivity as follows:

$$k_{eff} = k_f \left[1 + \frac{k_p \varphi \, r_f}{k_f (1 - \varphi) r_p} \right] \tag{5.14}$$

where φ, r_f, and r_p are nanoparticle volume fraction, liquid particle radius, and nanoparticle radius, respectively. The thermal conductivity of nanoparticle is calculated as:

$$k_p = c \, \overline{u}_p \tag{5.15}$$

where nanoparticle mean velocity, $\overline{u}_p$, is given by:

$$\overline{u}_p = \frac{2 \, k_b T}{\pi \, \mu \, d_p^2} \tag{5.16}$$

where T, μ, and d_p are fluid temperature, fluid viscosity, and nanoparticle diameter, respectively.

5.2 THERMOPHYSICAL PROPERTIES OF BIOFLUID

Blood consists of 40%–45% red blood cells or erythrocytes, white blood cells or leukocytes, and platelets or thrombocytes and the rest, 55%–60% by volume is plasma [17]. The blood thermophysical properties depend on different parameters such as age, temperature, and hematocrit. However, in living organisms in general, and in large mammals in particular, general properties do not alter significantly because all mentioned parameters are regulated [8]. The knowledge of biofluid and bio-nanofluid thermophysical properties (such as blood and blood with nanoparticles) becomes essential when flow and heat transfer study of blood in drug delivery and new cancer therapy is considered. Table 5.1 shows the thermophysical blood properties [18].

As an example the effective thermal conductivity of blood with Al_2O_3 nanoparticles suspension as a bio-nanofluid is presented later. The model uses a two-step model that is based on parallel mixture rule, thermal resistance concept, and Maxwell. First, a model

Table 5.1 The thermophysical blood properties

Blood		Number/mm³	Size μm	Cell volume (%)
Blood cells	Red blood cells (erythrocytes)	5×10^6	~8	97
	White blood cells (leukocytes)	5×10^3	~15	2
	Platelet (thrombocytes)	3×10^5	~3	1
Plasma	Contains 91% water and 9% proteins (their size is in order of nanometer)			

Blood cells: 55% of blood; Plasma: 45% of blood.

based on the parallel mixture rule and thermal resistance concept is used to predict the blood cell thermal conductivity, as discussed in more detail in Section 5.2.1. Then the Maxwell and Leong et al. equations [18] are used in two different ways to predict the thermal conductivity of bio-nanofluid (blood + Al$_2$O$_3$), as described in more detail in Section 5.2.2. Fig. 5.1 depicts the schematic model of a cubic vessel. The vessel is full of blood that is completely mixed with particles and a thermal resistance model is used to show the heat transfer in x–direction, as illustrated [19].

5.2.1 Calculation of Blood Cell Thermal Conductivity

Blood is a combination of plasma, red blood cells, white blood cells, and platelets, as provided in Table 5.1 [20]. There are experimental data for thermal conductivity of plasma and red blood cells as well as blood itself. However, the thermal conductivity of the white blood cells and platelets is not known. Therefore two models are proposed for calculation of blood cell thermal conductivity.

First model: In this model, thermal conductivity of blood, k_{blood}, is determined to be 0.4 W/m·K by the parallel mixture rule, which is as follows [21]:

$$k_{blood} = \varphi_{plasma}\, k_{plasma} + \varphi_{blood\ cells}\, k_{blood\ cells} \tag{5.17}$$

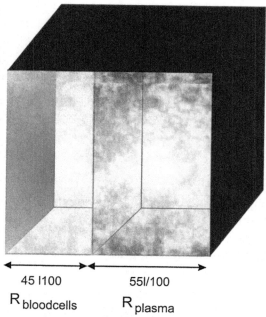

45 l100
$R_{bloodcells}$

55l/100
R_{plasma}

Figure 5.1 Thermal resistance in a cubic vessel full of blood.

Table 5.2 Plasma and blood thermal conductivity

	Volume fraction φ	Thermal conductivity (W/m·K)
Plasma	0.55	0.57 [8]
Blood cells	0.45	—
Blood	1	0.492 [8]
Blood	1	0.49—0.55 [18]

where all parameters except the blood cell thermal conductivity are known and are listed in Table 5.2 [22].

Second model: This model is based on thermal resistance concept [14]. As seen in Fig. 5.1, the model assumes a one-dimensional steady-state heat transfer and determines the thermal resistance between blood cells ($R_{blood\ cells}$) and plasma (R_{plasma}) as given below:

$$R_{blood} = R_{cells} + R_{plasma} \tag{5.18}$$

where thermal resistance of blood and plasma are as follows, respectively:

$$R_{bloodcell} = \frac{\varphi_{bloodcells} l}{k_{bloodcells} A} \tag{5.19}$$

$$R_{plasma} = \frac{\varphi_{plasma} l}{k_{plasma} A} \tag{5.20}$$

Therefore thermal conductivity of blood is determined by:

$$\frac{1}{k_{blood} A} = \frac{\varphi_{bloodcells} l}{k_{bloodcells} A} + \frac{\varphi_{plasma} l}{k_{plasma} A} \tag{5.21}$$

The values of all parameters in Eq. (5.21) are listed in Table 5.2 [22], with the only unknown being the blood cell conductivity. By using Eq. (5.21) and values listed in Table 5.2, the thermal conductivity of blood cells is determined to be 0.42 W/m·K.

5.2.2 Evaluation of Bio-Nanofluid Effective Thermal Conductivity

In this method, two approaches are used for evaluating the thermal conductivity of the bio-nanofluid mixture (blood + Al$_2$O$_3$ nanoparticles).

First Approach:

In this approach the Maxwell equation is used and the blood thermal conductivity, K_{eff}, is determined by:

$$K_{eff} = \frac{k_p + 2k_f + 2(k_p - k_f)\varphi}{k_p + 2k_f - (k_p - k_f)\varphi} k_f \tag{5.22}$$

Table 5.3 Fluid and particles data in Maxwell equation

Models	Fluid and particles	Volume fraction	Thermal conductivity (W/m·K)
First	Particles (blood cells)	0.45	0.4 (calculated)
	Fluid (plasma)	0.55	0.57
Second	Particles(blood cells)	0.45	0.4215 (calculated)
	Fluid (plasma)	0.55	0.57

where k_p is the particle thermal conductivity (blood cells evaluated from Section 5.2.1) and k_f is the plasma thermal conductivity. Using Maxwell equation, the data listed in Table 5.3, and value of 0.4 for blood cell thermal conductivity, the calculated value of the blood thermal conductivity is 0.488 W/m·K, which is about 0.9% lower than the experimental value of 0.492 reported by Holmes [22].

However, when the value of blood cell thermal conductivity is set equal to 0.4215 W/m·K, the blood thermal conductivity becomes 0.499 W/m·K, which is 1.6% higher than the experimental results of Holmes [22].

Finally, the result of Leong et al. is used to evaluate the thermal conductivity of bio–nanofluid, blood mixed with Al_2O_3, by [23]:

$$K_{eff} = \frac{(k_p - k_{layer})\phi\, K_{layer}\left(2\beta_1^3 - \beta_2^3 + 1\right)}{\beta_1^3(k_p + 2k_{layer}) - (k_p - k_{layer})\phi\left(\beta_1^3 + \beta_2^3 - 1\right)}$$
$$+ \frac{(k_p + 2k_{layer})\beta_1^3\left(\phi\beta_2^3(k_{layer} - k_f) + k_f\right)}{\beta_1^3(k_p + 2k_{layer}) - (k_p - k_{layer})\phi\left(\beta_1^3 + \beta_2^3 - 1\right)} \quad (5.23)$$

where $\beta_1 = 1 + \beta/2$ $\beta_2 = 1 + \beta$ $\beta = h/r$ $k_{layer} = 10*k_{bf}$.

Also, r and h are nanoparticle radius and nanolayer thickness, respectively. The blood thermal conductivity ($k_f = K_{blood}$) is either 0.488 W/m·K based on the first model or 0.499 W/m·K according to the second model. Also, the nanoparticle thermal conductivity is assumed to be equal to $k_p = k_{Al_2O_3} = 46$ W/m·K. Other parameters are determined by Eq. (5.23) except for nanolayer thermal conductivity, which is assumed to be $k_{layer} = 10*k_f$, as suggested by Leong et al. [23]. They assumed that nanoparticles are covered by nanolayer. The block diagram for evaluation of the bio–nanofluid thermal conductivity is provided in Fig. 5.2.

Second Approach:

In the second approach, first the Leong et al. relation for the mixture of nanoparticles and plasma is considered [23]. In this model, plasma is the base fluid (k_f) and Al_2O_3 is the nanoparticle (k_p). Then the Maxwell equation, as given by Eq. (5.22), is used to calculate the thermal conductivity of bio–nanofluid. In Eq. (5.22) k_p is the blood cell thermal conductivity (as described in Section 5.2.1) and k_f is the thermal conductivity of the mixture of plasma and Al_2O_3, which is calculated by Leong et al. relation [23]. The block diagram for the second approach is shown in Fig. 5.3.

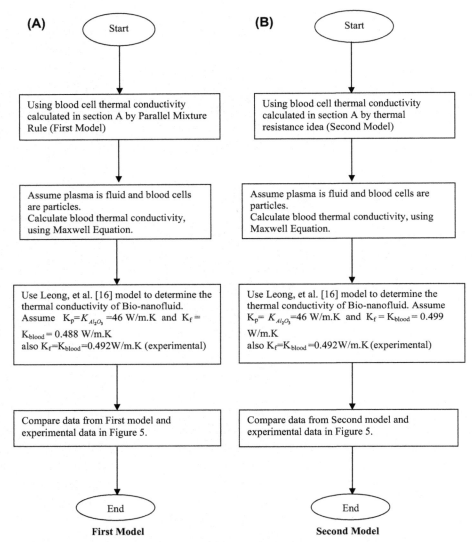

Figure 5.2 Block diagram for the first approach for evaluating the bio-nanofluid effective thermal conductivity (A) using first model; (B) using second model.

Fig. 5.4 depicts the comparison between predicted effective thermal conductivity of bio-nanofluid (blood with alumina nanoparticle) and that of the Leong et al. relation [23]. In our calculation of the nanolayer thermal conductivity, the thickness of the nanolayer is assumed to be 1 nm and the radius of nanoparticles is 5 nm. As shown, the predicted thermal conductivity values of the first and the second models are in good agreement with each other and with the earlier model given by Leong

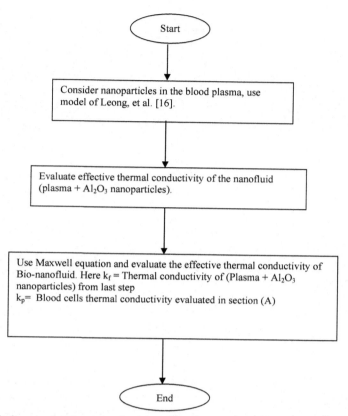

Figure 5.3 Block diagram for the second approach for evaluating the bio-nanofluid effective thermal conductivity.

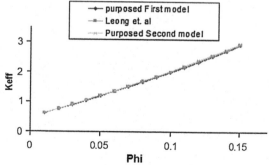

Figure 5.4 Effective thermal conductivity of bio-nanofluid versus volume fraction of Al_2O_3 nanoparticles. (Comparison of the proposed first and second models with that of Leong et al.)

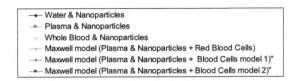

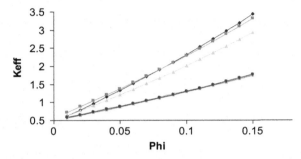

Figure 5.5 Effective thermal conductivity of bio-nanofluid versus volume fraction. (Comparisons of the proposed models with that of Leong et al.)

et al. [23]. In addition, the models also predict the expected increase in the value of nanofluid effective thermal conductivity as the volume fraction of nanoparticles increases; see Fig. 5.4.

Effective thermal conductivity of bio-nanofluid versus volume fraction is shown in Fig. 5.5. Thermal conductivity of bio-nanofluid, where the base fluid is the mixture of plasma and Al_2O_3 nanoparticles and blood cells are treated as microparticles, is determined by the second approach (Maxwell equation). In addition, the Leong et al. relation model is used for nanofluid with water as its base fluid [23]. Also, thermal conductivity of the blood cell was calculated using different methods and experimental value [24] was also used.

As shown in Fig. 5.5, nanofluid with water/plasma has the highest thermal conductivity, whereas blood thermal conductivity is the lowest. It is shown that the behavior of blood-based nanofluid thermal conductivity is similar to that of water-based nanofluid. As expected, the effective thermal conductivity of plasma that contains 91% water is very close to water thermal conductivity. The three predicted values of thermal conductivity that were calculated by the Maxwell model are almost identical. However, their value is lower than the one predicted by the first approach by a factor of 2. This is anticipated since Maxwell theory considers a lower thermal conductivity for the blood cells microparticles, whereas we assume that plasma size and nanoparticles are in the same order. Therefore it is fair to say that Maxwell theory predicts thermal conductivity of suspensions with microparticles more accurately than the other method.

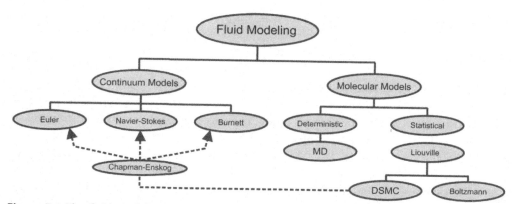

Figure 5.6 The fluid modeling classification. *DSMC*, direct simulation Monte Carlo; *MD*, molecular dynamics.

5.3 SIMULATION AND MODELING TOOLS FOR NANOFLUID

It is mandatory to consider the unconventional effects of nanofluid behavior such as slip flow, thermal creep, viscous dissipation, surface effect, and intermolecular forces in microsystem especially when selection of a proper fluid flow simulation method is needed. There are two basic tools for fluid flow modeling, continuum model and molecular model. Fig. 5.6 shows the classification of different fluid modeling.

A fixed amount of mass is used when the continuum model is utilized. For simulation purposes it is easier to use the continuum model compared with the molecular model. In the continuum model, the Navier–Stokes equations, the mass, momentum, and energy equations are typically used to simulate the flow as well as temperature field and the molecular nature of the gases or liquid is ignored. These coupled nonlinear Navier–Stokes and energy equations are generally solved using a finite difference or finite element computer codes. Even though the Navier–Stokes framework has long been established and used to predict the behavior of fluid, it is still unable to resolve some of the basic problems such as problems with large Knudsen number. Knudsen number is the ratio of mean free molecule path of the gases to their characteristic length. Such restrictions occur when the mean free path of the fluid molecules is similar to the geometrical system constraints. In many industrial or biomechanical applications, where the Knudsen number is typically less than 10^{-3}, the continuum model for flow is considered and accurate results are obtained, as noted in Section 3.2.

To numerically study and track the nanoparticles in base fluid the Eulerian and Lagrangian methods are used. In the Eulerian method, in which the fluid properties are written as function of time and space, the concentration equation in addition to momentum and energy equations is considered and solved. In Lagrangian approaches the particles are tagged and tracked individually as they move in the blood. Therefore the number of equations is equal to the number of particles. The interaction between each particle becomes important.

The most important forces between particles are magnetic force and hydrodynamic force. If a more realistic solution is intended, other forces such as weight, Brownian force, Basset force, lift force, and acoustical force should be considered.

The use of proper molecular models, for analyzing the behavior of some microsystems, is mandatory. In molecular models the fluid is considered as many discrete particles, molecules, atoms, ions, and electrons. The aim of the model is to determine the position, velocity, and state of all particles at all times. The molecular models are deterministic or probabilistic. The motions of the molecules are governed by the law of classical mechanics as well as quantum mechanics laws. Molecular Dynamics (MD) and Lattice-Boltzmann (LB) are the two most practical nonclassical model and simulation method.

MD has emerged as one of the first simulation methods in the late 1950s and early 1960s [25]. It has been widely used to study the structure and dynamics of macromolecules. Many microscopic and nanoscale problems are simulated by MD. Molecular dynamic simulation is also used in nanoscale systems due to the small size of its particles. For example, MD simulation is used to investigate the thermal resistance of a solid–fluid interface in presence of laminar shear flow [25]. Also, the flow of a liquid in a complex nanochannel is generally modeled by the MD model due to slip boundary condition in some of the flow features [26].

MD methods integrate Newton's equations of motion for a set of molecules on the basis of an intermolecular potential. In this method the equations of motion are discretized and solved numerically. MD simulation provides the time evolution of the interacting particles. Some numerical integrator, such as the Verlet algorithm, is used to determine the positions and velocities of particles within a finite time interval [27]. One of the most important steps in MD simulation is to determine the potentials between different particles. Potentials are either bond potentials or nonbond potentials. The most common potentials are electrostatic potentials, covalent bonds, and Lennard-Jones, LJ, forces. When two particles are far away from each other there exists no potential between them. Therefore a cut-off radius (r_c) is defined to reduce the computation cost. For LJ potential, the critical radius is determined by $r_c = (2.5-3)\sigma$ where σ is the bond length.

The next step is to determine the forces that act upon the particles. The force acting upon the i-th particle, $\boldsymbol{F}_i$, at time t is given by:

$$\boldsymbol{F}_i = m_i \frac{d^2 \boldsymbol{r}_i(t)}{dt^2} \tag{5.24}$$

where $\boldsymbol{r}_i(t)=(x_i(t),y_i(t),z_i(t))$ is the position vector of the i-th particle and m_i is the mass of the particle.

The lattice Boltzmann, LB, method is a powerful technique for the computational modeling. It solves the discrete Boltzmann equation instead of the Navier–Stokes equations. The LB method is derived from lattice gas automata, a discrete particle kinetics that utilizes a discrete lattice and a discrete time. The LB method is able to simulate a wide variety of problems such as hydrodynamic and magnetohydrodynamic (MHD) problems, multiphase and multicomponent fluids including suspensions and emulsions, chemical-reactive flows, and multicomponent flow through porous media [28].

The recently proposed Bhatnagar–Gross–Krook (BGK) Lattice Boltzmann kinetic scheme offers a promising tool for simulating complex three-dimensional MHD flows. It is based on pseudoparticle approaches, which are grouped into the lattice-based and off-lattice models. The lattice-based models include lattice gas method and LB method. The off-lattice models include the dissipative particle dynamics and the direct simulation Monte Carlo methods in conjunction with Newtonian dynamics.

The LB gas uses the particle velocity distribution function, $f_i(\overrightarrow{x_i}, t)$, to observe the pseudofluid particle with discrete velocity $\overrightarrow{c_i}$ at lattice node $\overrightarrow{x_i}$ at time t, as follows:

$$f_i^{new}(\overrightarrow{x} + \overrightarrow{c_i}\Delta t, t + \Delta t) - f_i^{old}(\overrightarrow{x}, t) = \Delta t \Omega_i \tag{5.25}$$

where index i, Δt, and Ω_i represent the n base vectors of the underlying lattice type, the lattice time step, and the collision operator, respectively. The left-hand term is the advection term and shows the free propagation of the particle packets along the lattice links. The collision operator, Ω_i, is determined by the BGK single-step relaxation approximation as [29]:

$$\Omega_i = -\frac{1}{\tau}\left(f_i^{old}(\overrightarrow{x}, t) - f_i^{eq}(\overrightarrow{x}, t)\right) \tag{5.26}$$

where τ is the relaxation time parameter that quantifies the rate of change toward local equilibrium for incompressible isothermal materials. In LB BGK method approximation all particle distributions are relaxed at the same rate, $\omega = \frac{1}{\tau}$, toward their corresponding equilibrium value. The equilibrium distribution, $f_i^{eq}(\overrightarrow{x}, t)$, depends only on locally conserved quantities such as mass density and momentum density [29].

In LB the macroscopic parameters such as local particle density, $\xi(\overrightarrow{x}, t)$, local mass density, $\rho(\overrightarrow{x}, t)$, local velocity vector, $\overrightarrow{u}(\overrightarrow{x}, t)$, momentum density vector, $\rho(\overrightarrow{x}, t)\overrightarrow{u}(\overrightarrow{x}, t)$, and local kinetic energy density, $\vartheta(\overrightarrow{x}, t)$, are determined by integrating the distribution functions as follows:

$$\xi(\overrightarrow{x}, t) = \sum_{i=0}^{n} f_i(\overrightarrow{x}, t) \tag{5.27}$$

$$\rho(\overrightarrow{x}, t) = m \sum_{i=0}^{n} f_i(\overrightarrow{x}, t) \tag{5.28}$$

$$\overrightarrow{u}(\overrightarrow{x}, t) = \frac{1}{\xi(\overrightarrow{x}, t)} \sum_{i=0}^{n} f_i(\overrightarrow{x}, t)\overrightarrow{c_i} \tag{5.29}$$

$$\rho(\overrightarrow{x}, t)\overrightarrow{u}(\overrightarrow{x}, t) = m\sum_{i=0}^{n}f_i(\overrightarrow{x}, t)\overrightarrow{c_i} \tag{5.30}$$

$$\vartheta(\overrightarrow{x}, t) = \frac{1}{2}\sum_{i=0}^{n}f_i(\overrightarrow{x}, t)\left|\overrightarrow{c_i} - \overrightarrow{u}(\overrightarrow{x}, t)\right|^2 \tag{5.31}$$

where $\overrightarrow{c_i}$ is the velocity vector. A popular way of classifying different lattice methods is the DnQm scheme. Here "Dn" stands for "n dimensions" whereas "Qm" stands for "m speeds." The most frequent mesh types for LB simulations are the D1Q3-, D2Q9-, D3Q15-, and the D3Q19-lattice [29].

Elementary and complex boundary conditions are the two most used conditions applied to the velocity field by the LB method. In elementary boundary condition, the physical boundary is aligned with the grid coordinates. The surface does not need to be smooth and it does not cut through mesh cells. Some of the elementary boundary conditions are: Periodic, No-slip, Free-slip, Frictional slip, Sliding walls, and Open inlet/outlet boundary conditions [30]. The complex boundary condition can take virtually any shape, including mesh-cutting surfaces. It includes stair casing, extrapolation, and surface elements dynamics [30].

5.4 BIOFLUID SIMULATION IN CARDIOVASCULAR SYSTEMS

The cardiovascular system (CVS) provides the nutrients, gases, and waste to and from the cells and heat through blood convection. It consists of heart, systemic circulation, pulmonary circulation, and nervous and biochemical regulators. Biochemical regulators modify the vessels' parameters to regulate the blood flow rate and pressure. The heart pumps blood throughout the systemic circulation. The aorta is the largest artery from the heart's left ventricle and extends down to abdomen, where it branches into smaller arteries. The first step of the cardiac cycle (a closed-loop, pulsatile system) is ventricular diastole. In this step the ventricles are relaxed and allow for newly oxygenated blood to flow in. When the ventricles contract and eject the blood out to the body through the aorta, the systole happens. The maximum pressure when blood is pumping into the aorta is systolic pressure. When the blood begins to flow into the ventricles, the aortic pressure is at the lowest, the diastolic pressure.

Computational tools are means that help the understanding of blood behavior in an inexpensive and noninvasive way. Several different mathematical models have been developed to simulate the biofluid in CVS since 1899 [31]. These models provide better diagnostics and physiological understanding, cardiac prosthesis, and medical planning. Modern technology tools such as computer-based simulations and image technology have eased the understanding of the complex nature of interactions and physiological functions of the CVS.

Table 5.4 The relation between fluid dynamics, electrical circuit, and blood flow circulation

Physiological variables	Fluid dynamics	Electrical analog
Blood pressure (mmHg)	Pressure P (J/m^3)	Voltage U (volt, J/C)
Blood flow rate (L/s)	Flow rate Q (m^3/s)	Current I (ampere, C/s)
Blood volume (L)	Volume V (m^3)	Charge q (C)
Blood resistance R	Viscosity μ	Electrical resistance R
Vessel's wall compliance	Elastic coefficient	Capacitor's capacity (C)
Blood inertia	Inertance	Inductor's inertance L
$Q = \Delta P/\Delta R = \Delta P \pi r^4/8 \, \mu L$	Poiseuille's law	Ohm's law $I = \Delta U/R$

5.4.1 Lumped Model

The lumped model, also known as 0-D model, was initially introduced by Joseph C. Greenfield and Donald L. Fry [31]. This model provides the global distribution of the pressure, flow rate, and blood volume for specific physiological conditions. However, in systemic arterial network the whole cardiovascular dynamics are assumed as a closed hydraulic circuit. This lumped model is known as Windkessel (WK) model and was first introduced by Otto Frank [32]. The WK model uses resistance, compliance, and impedance to describe the hemodynamics of the arterial system. Impedance is the ratio of wave speed multiplied by blood density to aortic cross-sectional area. The blood flow circulation in human body is generally modeled by electrical circuit. Table 5.4 shows the relation between fluid dynamics, electrical circuit, and blood flow circulation in the human body.

The simplest form of the WK models is the two–element model, a model with a resistance and a compliance element [33]. The WK model ignores the veins and uses a capacitor C and a resistor R to describe the depository properties of large arteries as well as the dissipative nature of peripheral vessels.

The theoretical two–element WK model is [34]:

$$Q(t) = \frac{P(t)}{R} + C \, \frac{dP(t)}{dt} \qquad (5.32)$$

where $P(t)$ in mmHg and $Q(t)$ in cm^3/s are the blood pressure in the aorta and flow of blood from the heart. Also, C in cm^3/mmHg and R in mmHg s/cm^3 are the arterial compliance and the peripheral resistance of the systemic arterial system.

The peripheral resistance, R, is inversely proportional to blood vessel radius to the fourth power, as stated by Poiseuille's law and is calculated by:

$$R = \left(p_{ao,mean} - p_{ven,mean}\right)/co \approx p_{ao,mean}/co \qquad (5.33)$$

where $p_{ao,mean}$, $p_{ven,mean}$, and CO are mean aortic pressure, mean venous pressure, and cardiac output, respectively. The compliant element is mainly determined by the elasticity of the large, or conduit, arteries. The total arterial compliance, C, is the ratio of a volume change, DV, to the resulting pressure change DP.

To improve the result, Windkessel introduced a three-element model. The three-element WK model is developed by inserting the additional resistor ZC into the two-element model. Impedance ZC represents the aortic characteristic impedance of the arterial tree with dimension that resembles resistance. The three-element WK model is largely used in CVS simulation. It improves the simulation of the high-frequency components by predicting the diastolic and systolic aortic pressures accurately [35]. In this model the ZC element is parallel with capacity C. The tree-element WK model is given by [34]:

$$\left(1 + \frac{ZC}{R}\right)Q(t) + CR_1\frac{dQ(t)}{dt} = \frac{P(t)}{R} + C\,\frac{dP(t)}{dt} \tag{5.34}$$

The three-element WK model describes the pressure-flow relations at the entrance of the arterial system. This model is used for the systemic arterial and the pulmonary arterial of human and other mammals. The three-element WK model is easier to use compared with distributed models [36]. However, in a lumped model, limited number of meaningful physiological parameter such as local vascular changes are considered.

Another three-element WK model was proposed by Buratini [37]. He placed a small resistor ZC in series with the capacitor C, which shows the elastoviscous effect of the vessel's wall. This is different from the ZC element introduced by Westerhof who describes the wave reflection during the flow.

To improve the WK model, Stergiopulos [38] introduced the four-element WK model by adding inductive element, L. Inductive element describes the internal effect of the blood flow. The improved four-element WK model is given as follows [34]:

$$\left(1 + \frac{ZC}{R}\right)Q(t) + \left(ZC \times C + \frac{L}{R}\right)\frac{dQ(t)}{dt} + LC\frac{d^2Q(t)}{dt^2} = \frac{P(t)}{R} + C\frac{dP(t)}{dt} \tag{5.35}$$

Another four-element WK model was introduced by Grant [39]. He added the inductive element into the Buratini model. This model is suitable at high frequencies of arterial input impedance. Fig. 5.7 shows the schematic of different types of WK models.

5.4.2 One-Dimensional Model

The lumped model does not consider the pressure and propagation wave in arterial tree. To take into consideration the wave propagation and vessel geometry blood flow simulation, one-dimensional model based on Euler equations is used. In this model the vessel is assumed as a cylinder and the blood flows along its axis. Blood is assumed to be a Newtonian fluid. The mass and momentum equations for one-dimensional impermeable deformable tubular control volume are formulated as:

$$\frac{\partial U}{\partial t} + H\frac{\partial U}{\partial x} = S \tag{5.36}$$

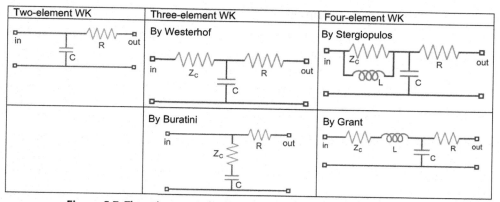

Figure 5.7 The schematic of different types of Windkessel (WK) models.

$$U = \begin{bmatrix} A \\ U \end{bmatrix} \quad H = \begin{bmatrix} U & A \\ \dfrac{1}{\rho}\dfrac{\partial P}{\partial A} & U \end{bmatrix} \quad S = \begin{bmatrix} 0 \\ \dfrac{1}{\rho}\left(\dfrac{f}{A} - S\right) \end{bmatrix} \tag{5.37}$$

where x is the axial coordinate along the vessel, t is time, S is the source term, $A(x,t)$ is the lumen cross-sectional area, $U(x,t)$ is the average axial velocity, and $P(x,t)$ is the average internal pressure over the cross-section. In Eq. (5.37) f is shear stress and is determined by:

$$f = 2\mu\pi\widehat{R}\left[\frac{\partial u}{\partial r}\right]_{r=\hat{R}} \tag{5.38}$$

where $\widehat{R}(x,t)$ and $u(x,r,t)$ are the lumen radius and the axial velocity, respectively. A typical profile for no-slip velocity, u, boundary condition is:

$$u = U\frac{\gamma+2}{\gamma}\left[1 - \left(\frac{r}{R}\right)^{\gamma}\right] \tag{5.39}$$

where γ is constant. Setting $\gamma = 9$ provides close value to the experimental data. For Poiseuille's flow $\gamma = 2$ and $f = -8\mu\pi U$. The pressure—area relation is given by [40]:

$$P = \frac{\beta}{A_0}\left(\sqrt{A} - \sqrt{A_0}\right) \quad , \quad \beta(x) = \frac{4}{3}\sqrt{\pi}\ hE \tag{5.40}$$

where $h(x)$, $E(x)$, and $A_0(x)$ are the thickness, Young's modulus, and a lumen area at the reference state $(P,U) = (0,0)$, respectively. Source term, S, is given by [40]:

$$S = \frac{\partial P}{\partial \beta}\frac{d\beta}{dx} - \frac{\partial P}{\partial A_0}\frac{dA_0}{dx} \tag{5.41}$$

5.4.3 Two- and Three-Dimensional Simulation

Generally one-dimensional models do not provide enough details that are required to understand the behavior of the blood flow in human body. Therefore most researchers use either two- or three-dimensional models when simulating blood flow in arterial tree. In the two-dimensional model the geometry of human vessels is depicted as either axisymmetric or planner, whereas in three-dimensional model it is important to use geometry that is identical to human arterial tree and vessels. In both cases the Navier—Stokes equations and related boundary conditions are applied and solved by either commercial software or in-house codes to help the understanding of the blood flow behavior in human body. Among all, finite element- and finite volume-based methods are used the most.

Sometimes it is necessary to consider the interaction between blood flow and walls of the vessel and artery. These types of problems are referred to as fluid solid interaction model where two domains, solid and fluid, are considered. The solid and fluid domains are solved with Arbitrary Lagrangian-Eulerian method.

5.5 SIMULATION OF FLUID FLOW AND HEAT TRANSFER OF BIO-NANOFLUID

Application of nanotechnology in medicine and cancer therapy has generated a lot of interest in investigation of bio-nanofluid flow such as blood with nanoparticles. The thermophysical properties are necessary for flow and heat transfer study. The aim of this section is to investigate the behavior of bio-nanofluid flow, such as blood flow that is mixed with nanoparticles, and heat transfer. The thermophysical properties of blood mixed with Al_2O_3 nanoparticles were numerically determined by Shahidian et al. [18]. After that a computational fluid dynamic code based on finite volume is used to solve the governing continuity and momentum equations for the bio-nanofluid flow in two-dimensional channels. Blood and Al_2O_3 nanoparticles thermophysical properties are given in Table 5.5.

Effective thermal conductivity of bio-nanofluids depends on their size and blood cells and plasma properties. Initially Leong et al. [23] introduced a relation that does not take

Table 5.5 Thermophysical properties of blood and Al_2O_3 nanoparticles

Property	Human blood	Al_2O_3 nanoparticles
Density (kg/m^3)	1060	3900
Thermal conductivity (W/m·K)	0.492	35
Viscosity (cp)	3	—
Specific heat capacity (J/kg·°C)	3750	775

into account blood cells and assumes plasma as base fluid (k_f) and Al_2O_3 as nanoparticles (k_p). They assumed that nanoparticles are covered by nanolayer with thermal conductivity equal to $K_{nanolayer} = 10 * K_f$.

Then an improved bio-nanofluid thermal conductivity relation that contains plasma, nanoparticles, and blood cells was introduced by Maxwell [3]. The proposed relation by Maxwell provides good estimate when applied to the fluid with suspended microparticles. Combining Maxwell's equation Eq. (5.22) with the blood cells as particles (k_p) and mixture of plasma and Al_2O_3 as base fluid (k_f) gives effective thermal conductivity of bio-nanofluid, K_{eff}. The effective density, specific heat capacity, and viscosity of bio-nanofluid are determined by Eqs. (5.1–5.3) respectively.

The nonlinear governing continuity, momentum, and energy equations for steady, laminar, incompressible, and Newtonian fluid are given by the following equations, respectively:

Continuity:

$$\frac{\partial u}{\partial x} + \frac{\partial v}{\partial y} = 0 \tag{5.42}$$

X—Momentum:

$$u\frac{\partial u}{\partial x} + v\frac{\partial u}{\partial y} = -\frac{\partial p}{\partial x} + \mu\left(\frac{\partial^2 u}{\partial x^2} + \frac{\partial^2 u}{\partial y^2}\right) \tag{5.43}$$

Y—Momentum:

$$u\frac{\partial v}{\partial x} + v\frac{\partial v}{\partial y} = -\frac{\partial p}{\partial y} + \mu\left(\frac{\partial^2 v}{\partial x^2} + \frac{\partial^2 v}{\partial y^2}\right) \tag{5.44}$$

Energy equation:

$$\rho C_p\left(\frac{\partial T}{\partial t} + \overrightarrow{V} \cdot \boldsymbol{\nabla} T\right) = k\boldsymbol{\nabla}^2 T \tag{5.45}$$

Boundary conditions: the inlet velocity is assumed to be 0.05 m/s and at the channel outlet the first derivative of velocity is zero.

Fig. 5.8 depicts the bio-nanofluid effective thermal conductivity for different volume fractions of Al_2O_3 nanoparticles.

As expected the bio-nanofluid effective thermal conductivity changes for low volume fraction of nanoparticle (less than 1%) is small.

The blood velocity and bio-nanofluid (blood with 0.1% Al_2O_3 nanoparticle) velocity contour are shown in Fig. 5.9.

As shown there is no significant difference between blood velocity contour and bio-nanofluid velocity contour.

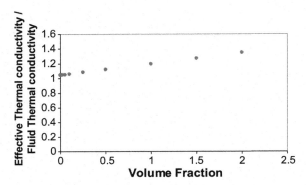

Figure 5.8 The bio-nanofluid effective thermal conductivity.

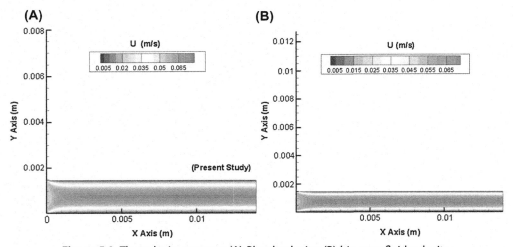

Figure 5.9 The velocity contour. (A) Blood velocity; (B) bio-nanofluid velocity.

Velocity distribution for different volume fractions of Al_2O_3 nanoparticle at $y = 0.00036$ m from the bottom edge of the channel is shown in Fig. 5.10.

Also, Fig. 5.11 depicts velocity distribution for different volume fractions of Al_2O_3 nanoparticle as a part of Fig. 5.10 in large scale.

The heat transfer of bio-nanofluid with different nanoparticle (Al_2O_3) volume fraction is investigated. The bio-nanofluid flow in microchannel with 10,000 W/m^2 heat flux on the wall (boundary layer) is studied. Fig. 5.12 shows the temperature at 0.00036 m distance from the bottom of microchannel for different nanoparticles volume fraction.

Fig. 5.13 shows the temperature at 0.00036 m distance from the bottom of microchannel for different nanoparticles volume fraction in large scale.

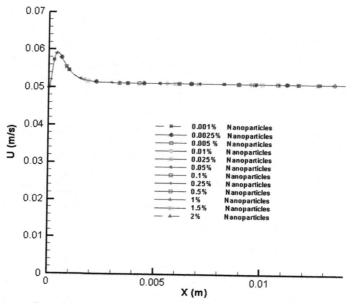

Figure 5.10 Bio-nanofluid velocity distribution in channel for different volume fractions of Al_2O_3 nanoparticle.

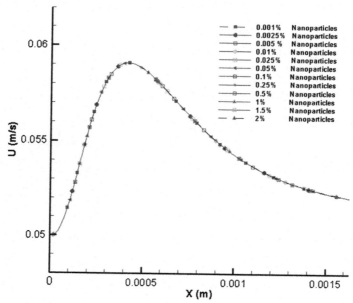

Figure 5.11 Large-scale bio-nanofluid velocity distribution for different volume fractions of Al_2O_3 nanoparticle.

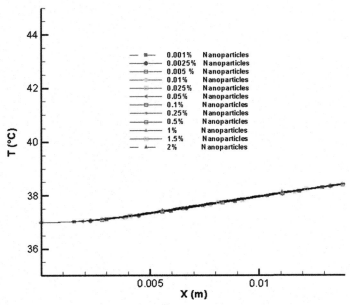

Figure 5.12 Bio-nanofluid temperature at 0.00036 m distance from the bottom of microchannel for different nanoparticles volume fraction, with 10,000 W/m² heat flux on the wall.

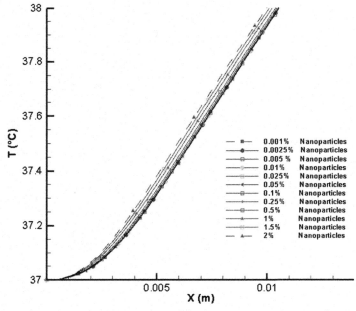

Figure 5.13 Large-scale bio-nanofluid temperature at 0.00036 m distance from the bottom of microchannel for different nanoparticles volume fraction, with 10,000 W/m² heat flux on the wall.

The simulation results show that the thermophysical properties increase. Flow, velocity, and pressure of blood and bio-nanofluid are the same. As shown in the above-mentioned figures, as nanoparticle volume fraction changes from 0.001% to 2%, the velocity stays constant inside the channel while the temperature increases from 37.93 to 38°C. This means that it is safe to use nanoparticles for drug delivery in biomedical application. This is an important and useful result in nano drug delivery applications.

REFERENCES

[1] X. Qi Wang, S. Mujumdar, Heat transfer characteristics of nanofluids: a review, International Journal of Thermal Sciences 46 (2007) 1−19.

[2] J. Koo, C. Kleinstreuer, A new thermal conductivity model for nanofluids, Journal of Nanoparticle Research 6 (2004) 577−588.

[3] J. Maxwell, A Treatise on Electricity and Magnetism, second ed., Oxford University Press, Cambridge, UK, 1904.

[4] R. Hamilton, O. Crosser, Thermal conductivity of heterogeneous two-component systems, I & EC Fundamentals 125 (3) (1962) 187−191.

[5] Q.-Z. Xue, Model for effective thermal conductivity of nanofluids, Physics Letters A 307 (2003) 313−317.

[6] W. Yu, S.U.S. Choi, The role of interfacial layers in the enhanced thermal conductivity of nanofluids: a renovated Maxwell model, Journal of Nanoparticle Research 5 (1−2) (2003) 167−171.

[7] W. Yu, S.U.S. Choi, The role of interfacial layers in the enhanced thermal conductivity of nanofluids: a renovated Hamilton-crosser model, Journal of Nanoparticle Research 6 (4) (2004) 355−361.

[8] S.P. Jang, S.U.S. Choi, Applied Physics Letters 85 (2004) 3549.

[9] R. Prasher, P. Bhattacharya, P.E. Phelan, Thermal conductivity of nanoscale colloidal solutions (nanofluids), Physical Review Letters 94 (2) (2005) 025901.

[10] P. Keblinski, S. Philipot, S. Choi, J. Eastman, Mechanism of heat flow in suspensions of nano-sized particles (nanofluids), International Journal of Heat Mass Transfer 45 (2002) 855−863.

[11] P. Keblinski, J.A. Eastman, D.G. Cahill, Nanofluids for thermal transport, Materials Today 8 (6) (2005) 36−44.

[12] A. Shahidian, Flow and Thermal Analysis of Bio-nano Fluid in a Channel Like Vessel under the Effect of Electromagnetic Forces (MHD), Submitted in partial fulfillment of the requirements for the PhD degree in Mechanical Engineering, K.N. Toosi University of Technology, January 2011.

[13] L.E. Neilsen, Predicting the Properties of Mixtures: Mixture Rules in Science and Engineering, Marcel Dekker, New York, 1978.

[14] O. Wiener, Die Theorie des Mischkörpers für das Feld der stationären Strömung: 1. Abhandlung: Die Mittelwertsätze für Kraft, Polarisation und Energie, Abh Math-Phys Kl Koniglich Saechsis Ges Wiss 32 (1912) 507−604.

[15] Y. Benveniste, Effective thermal conductivity of composites with a thermal contact resistance between the constituents: nondilute case, Journal of Applied Physics 61 (1987) 2840−2843.

[16] D.H. Kumar, H.E. Patel, V.R.R. Kumar, T. Sundararajan, T. Pradeep, S.K. Das, Model for heat conduction in nanofluids, Physics Review Letters 93 (14) (2004) 144301.

[17] J.N. Mazumdar, Biofluid Mechanics, World Scientific Publishing, 1992.

[18] M. Ghassemi, A. Shahidian, G. Ahmadi, S. Hamian, A new effective thermal conductivity model for a bio-nanofluid (blood with nanoparticle Al_2O_3), International Communications in Heat and Mass Transfer 37 (2010) 929−934.

[19] F.P. Incropera, D.P. Dewitt, Introduction to Heat Transfer, fourth ed., John Wiley & Sons, 2002.

[20] M. Thiriet, Biology and Mechanics of Blood Flows, Part I: Biology, Project-Team INRIA-UPMC-CNRS REO, Laboratoire Jacques-Louis Lions, CNRS UMR 7598, Université Pierre et Marie Curie, Springer Science, 2008.

[21] S.K. Das, S.U.S. Choi, W. Yu, T. Pradeep, Nanofluids Science and Technology, John Wiley & Sons, 2008.

[22] F. Kreith, D.Y. Goswami, B.I. Sandor, Part 4 of the CRC Handbook of Mechanical Engineering: Appendix a "Thermal Conductivity Data for Specific Tissues and Organs for Humans and Other Mammalian Species" by Kenneth. R. Holmes, CRC Press, 2004.

[23] K.C. Leong, C. Yang, S.M.S. Murshed, A model of the thermal conductivity of nanofluids—the effect of interfacial layer, Journal of Nanoparticle Research 8 (2006) 245—254.

[24] V.M. Nahirnyak, S.W. Yoon, C.K. Holland, Acousto-mechanical and thermal properties of sclotted blood, Journal of the Acoustical Society of America 119 (6) (2006) 3766—3772.

[25] D.C. Rapaport, The Art of Molecular Dynamics Simulation, Cambridge University Press, 2003.

[26] Xi-J. Fan, N. Pahn-Thien, N.T. Yong, D. Xu, MD simulation of a liquid in a complex nano channel flow, Physics of Fluids 14 (3) (March 2002).

[27] M. Allen, Introduction to MDS, J.V. Newman Institute for Computing, 2004.

[28] X. He, Li-S. Luo, Theory of the lattice Boltzmann method: from the Boltzmann equation to the lattice Boltzmann equation, Physical Review E 56 (1997).

[29] D.A. Wolf-Gladrow, Lattice-Gas Cellular Automata and Lattice Boltzmann Models (Lecture Notes in Mathematics, No. 1725), Springer, Berlin, 2000.

[30] S. Succi, The Lattice Boltzmann Equation: For Fluid Dynamics and beyond (Series Numerical Mathematics and Scientific Computation), Oxford University Press, Oxford, 2001.

[31] L. Waite, J. Fine, Applied Biofluid Mechanics, Mc-Graw Hill, 2007, http://dx.doi.org/10.1036/0071472177.

[32] I. Kokalari, T. Karaja, M. Guerrisi, Review on lumped parameter method for modeling the blood flow in systemic arteries, Journal of Biomedical Science and Engineering 6 (2013) 92—99.

[33] O. Frank, Die grundform des arterielen pulses erste abhandlung: mathematische analyse, Zeitschrift für Biologie 37 (1899) 483—526.

[34] M. Catanho, M. Sinha, V. Vijayan, Model of Aortic Blood Flow Using the Windkessel Effect, BENG 221-Mathematical Methods in Bioengineering Report, October 25, 2012.

[35] D. Burkhoff, J. Alexander, J. Schipke, Assessment of Windkessel as a model of aortic input impedance, American Journal of Physiology—Heart and Circulatory Physiology 255 (1988) 742—753.

[36] N. Westerhof, J.-W. Lankhaar, B.E. Westerhof, The arterial Windkessel, Medical & Biological Engineering & Computing (2008), http://dx.doi.org/10.1007/s11517-008-0359-2. Special Issue—Review.

[37] R. Burattini, S. Natalucci, Complex and frequency-dependent compliance of viscoelastic Windkessel resolves contradictions in elastic Windkessels, Medical Engineering & Physics 20 (1998) 502—514.

[38] N. Stergiopulos, B.E. Westerhof, N. Westerhof, Total arterial inertance as the fourth element of the Windkessel model, American Journal of Physiology—Heart and Circulatory Physiology 276 (1992) 81—88.

[39] B.J.B. Grant, L.J. Paradowski, Characterization of pulmonary arterial input impedance with lumped parameter models, American Journal of Physiology—Heart and Circulatory Physiology 252 (1987) 585—593.

[40] J. Alastruey, K.H. Parker, J. Peiro, S.J. Sherwin, Lumped parameter outflow models for 1-D blood flow simulations: effect on pulse waves and parameter estimation, Global Science (2008) 317—336.

CHAPTER 6

Nanosystem Application in Drug Delivery

The idea of drug delivery by nanocarriers as a novel concept in medicine is still in its primary stages. The ultimate goal of engineers is to design nanosystems that can carry and deliver drug to specific human organs or cells with minimal side effects and toxicity without affecting other parts of the human body. One potential problem with nanocarriers may arise from the unwanted toxicity from the inorganic nanomaterial, which is toxic to the human body if it accumulates in certain cells or organs.

The most promising use of nanocarriers is in chemotherapy. As known, drugs used in chemotherapy are extremely toxic. It is therefore vital to deliver and release them to the tumor and avoid other parts of the human body.

In the following section the behavior of a nanoparticle as a drug carrier inside the human body is discussed. The governing equations are introduced and an efficient path to deliver drug is determined in such a way that the side effects are minimized.

6.1 INTRODUCTION

Drug delivery systems (DDS) are systems based on interdisciplinary sciences such as polymer science, pharmaceutics, bioconjugated chemistry, and molecular biology. DDS transport pharmaceutical compounds by nanocarriers to the desired location inside the human body in a safe manner. The goal of the nanocarriers is to deliver and release the drug in a manner that maximizes the efficiency and minimizes the side effect. The delivery of drug into human body is designed based on several factors such as disease types, desired effect, and product availability. To achieve efficient targeted delivery, the designed system must avoid the host defense mechanisms and circulate to its intended site of action [1]. In targeted delivery (i.e., delivery of drug to cancerous tissues) the drug is released over a period of time and in a controlled manner to achieve best efficiency.

Among different anatomical routes by which drugs are introduced into the human body, the noninvasive perioral (through the mouth), topical (skin), transmucosal (nasal, buccal/sublingual, vaginal, ocular, and rectal), and inhalation routes are the most common ones. The only concern with these routes, however, is that many medications, such as antibodies, peptides, proteins, vaccines, and gene-based drugs, cannot be delivered to the location appropriately. This is due to molecular size and charge issues that make these medications susceptible to enzymatic degradation and they cannot be

Nano and Bio Heat Transfer and Fluid Flow
ISBN 978-0-12-803779-9, http://dx.doi.org/10.1016/B978-0-12-803779-9.00006-6

absorbed by the body. Therefore injection or nanoneedle array is used for the delivery of many protein and peptide drugs.

6.1.1 Nanodrug Carrier

Nanodrug carriers are the newest tools in drug delivery. Nanocarriers are of size 1—100 nm and can deliver drugs to sites that are otherwise inaccessible in the human body. They are classified based on: material, shape, and different medical applications. Nanocarriers are made from materials such as lipid-based materials (i.e., micelles and liposomes), polymeric nanoparticles, carbon-based materials (i.e., nanotubes), and gold nanoparticles (i.e., nanoshells and nanocages). The micelles are used to carry hydrophobic and hydrophilic drugs effectively inside the human body. In addition, protein-based nanocarriers are used therapeutically since they occur naturally and generally demonstrate less cytotoxicity than synthetic molecules.

Nanocarriers are designed to have high material binding and drug retention capacity. The drugs are either loaded into or encapsulated inside the carriers by emulsion technique first and then are transported to the desire location in a controlled manner. They disperse the drug into the target homogenously and release the drug at a predetermined rate for a required period of time. They not only do not get damaged from time of attachment to the carrier until it has been delivered but also avoid damage to drug structure when the drug is incorporated into the carrier. To achieve effective drug delivery, nanocarriers must have suitable circulation time to reach their target. Some important key factors that impact the efficiency of drug delivery systems are size, shape, and surface characteristics of nanocarriers. Details of different nanocarriers including nanoparticle size range, type of therapeutic drug they carry, and specific advantages and limitations are given in Table 6.1 [2].

Particle size plays a major role in particle functions, such as degradation, vascular dynamics, targeting, and clearance and uptake mechanisms. In addition, the shape of particles has an interesting effect on particle functions, their transportation through the blood vessels, and targeting unhealthy sites. Another important factor is surface characteristics of nanocarriers, which determine their lifespan during circulation in the blood stream.

Drug delivery by nanocarriers to the specific location is achieved passively (outer surface of nanocarriers is coated with polyethylene oxide), actively (outer surface of nanocarriers is coated with ligands or antibodies), by releasing drugs in specific pH ranges (micelle-polymer nanocarriers), or by releasing drugs at certain temperatures.

Drug release occurs by diffusion of biomolecules through pores, degradation of carriers, or swelling of materials in carriers. The release of drug from carrier materials depends on physiochemical properties of polymer and drug, and physiological properties of release area.

Table 6.1 Nanocarrier with various materials

Nanocarrier	Properties
Nanocrystals	Synthesized by the reprecipitation method, the resulting drug nanocrystals were stable in aqueous dispersion, without the necessity of any additional stabilizer. These nanocrystals are uniform in size distribution with an average diameter of 110 nm. These nanocrystal drugs may have advantages over association colloids (micelle solutions) because the level of surfactant per amount of drug can be greatly minimized, using only the amount that is necessary to stabilize the solid–fluid interface
Organic types:	
Liposomes	Liposomes are self-assembled artificial vesicles developed from amphiphilic phospholipids. These vesicles consist of a spherical bilayer structure surrounding an aqueous core domain, and their size can vary from 50 nm to several micrometers. Liposomes have attractive biological properties. Liposomes are the most clinically established nanosystems for drug delivery. Their efficacy has been demonstrated in reducing systemic effects and toxicity, as well as in attenuating drug clearance
Polymeric nanoparticles (NPs)	Polymeric NPs are colloidal particles with a size range of 10—1000 nm, and they can be spherical, branched, or core—shell structures. They have been fabricated using biodegradable synthetic polymers, such as polylactide—polyglycolide copolymers, polyacrylates, and polycaprolactones, or natural polymers, such as albumin, gelatin, alginate, collagen, and chitosan. Polymeric nanocarriers can be categorized based on three drug-incorporation mechanisms, polymeric carriers that use covalent chemistry for direct drug conjugation, hydrophobic interactions between drugs, and nanocarriers or hydrogels, which offer a water-filled depot for hydrophilic drug encapsulation
Polymer—drug conjugates (prodrugs)	Many polymer—drug conjugates have been developed since the first combination reported in the 1970s. Conjugation of macromolecular polymers to drugs can significantly enhance the blood circulation time of the drugs. Especially, protein or peptide drugs, which can be readily digested inside the human body, can maintain their activity by conjugation of the water-soluble polymer polyethylene glycol (PEG; PEGylation). In addition, polymer—drug conjugates are still limited by their nonbiodegradability and the fate of polymers after in vivo administration

Continued

Table 6.1 Nanocarrier with various materials—cont'd

Nanocarrier	Properties
Polymeric Micelles	Polymeric micelles are formed when amphiphilic surfactants or polymeric molecules spontaneously associate in aqueous medium to form core–shell structures. The inner core of a micelle, which is hydrophobic, is surrounded by a shell of hydrophilic polymers, such as PEG. Their hydrophobic core serves as a reservoir for poorly water-soluble and amphiphilic drugs; at the same time, their hydrophilic shell stabilizes the core, prolongs circulation time in blood, and increases accumulation in tumor tissues. So far, a large variety of drug molecules have been incorporated into polymeric micelles. Furthermore, using computer simulation, the experimental preparation of drug-loaded polymeric micelles could be more efficiently guided, by providing insight into the mechanism of mesoscopic structures and serving as a complement to experiments
Hydrogel NPs	Hydrogel NPs. In recent years, considerable attention has been paid to hydrogel NPs, which due to their unique features are considered as one of the most beneficial nanoparticulate drug delivery systems with unique properties. Swelling properties, network structure, permeability, or mechanical stability of hydrogels can be controlled by external stimuli or physiological parameters. Hydrogels have been extensively studied for controlled release of therapeutics, stimuli-responsive release, and applications in biological implants. Although hydrogel NPs-based drugs are not commercially available, they have high possibility to be further developed for drug delivery systems in the future, owing to their highly biocompatible and effective drug-loading properties
Protein–based NPs	Hydrophobic drugs, such as taxanes, are highly active and widely used in a variety of solid tumor therapies. Both paclitaxel (PTX) and docetaxel, which are the commercially available taxanes for clinical treatments, are hydrophobic. Because of their solubility problems, they have been formulated as suspensions with nonionic surfactants. However, these surfactants are associated with hypersensitivity reaction and toxic side effects to tissues

	Dendrimers	Dendrimers are synthetic, branched macromolecules that form a tree-like structure. Unlike most linear polymers, the chemical composition and molecular weight of dendrimers can be precisely controlled; hence, it is relatively easy to predict their biocompatibility and pharmacokinetics. Dendrimers are very uniform with extremely low polydispersities, and they are commonly created with dimensions incrementally grown in approximate nanometer steps from 1 to over 10 nm. Drug molecules associated with dendrimers can be utilized for cancer treatment. Since the clinical experience with dendrimers has so far been limited, it is hard to tell whether the dendrimers are intrinsically "safe" or "toxic."
Inorganic types:	Au NPs	Noble metal NPs, such as Au NPs, have emerged as a promising scaffold for drug and gene delivery in that they provide a useful complement to more traditional delivery vehicles. The combination of inertness and low toxicity, easy synthesis, very large surface area, well-established surface functionalization (generally through thiol linkages), and tunable stability provide Au NPs with unique attributes to enable new delivery strategies. Once the Au NPs are targeted to the diseased site, such as a tumor, hyperthermia treatment can be used for tumor destruction. The key issue that needs to be addressed with Au NPs is the engineering of the particle surface for optimized properties
	Superparamagnetic NPs	Magnetic NPs have been proposed as drug carriers with a push toward clinical trials. To guide microcapsules in place for delivery by external magnetic fields, the superparamagnetic features of iron (II) oxide particles can be used. The ability to heat the particles after internalization also known as the hyperthermia effect is another privilege of using magnetic NPs. Magnetic NPs are used for targeting and raising temperature. Moreover, the permeability of microcapsules can be affected by using oscillating magnetic fields that are external and releasing the materials that are encapsulated. The main benefits of superparamagnetic NPs over classical cancer therapies are minimal invasiveness, accessibility of hidden tumors, and minimal side effects. Conventional heating of a tissue by, for example, microwaves or laser light results in the destruction of healthy tissue surrounding the tumor. However, targeted paramagnetic particles provide a powerful strategy for localized heating of cancerous cells

Continued

Table 6.1 Nanocarrier with various materials—cont'd

Nanocarrier	Properties
Ceramic NPs	Ceramic NPs are particles fabricated from inorganic compounds with porous characteristics, such as silica, alumina, and titania. Among these, silica NPs have attracted much research attention as a result of their biocompatibility and ease of synthesis, as well as surface modification. For further in vivo applications, the biocompatibility, biodistribution, retention, degradation, and clearance of mesoporous silica nanoparticles (MSNs) must be systematically investigated
Carbon-based nanomaterials	Carbon-based nanomaterials have attracted particular interest because they can be surface functionalized for the grafting of nucleic acids, peptides, and proteins. Carbon nanotubes, fullerene, and nanodiamonds have been extensively studied for drug delivery applications. The size, geometry, and surface characteristics of single-wall nanotubes, multiwall nanotubes, and C_{60} fullerenes make them appealing for drug carrier usage. The toxicity of other forms of nanocarbons has also been reported. Given the mounting evidence demonstrating the toxicity of carbon NPs, the enthusiasm to develop carbon NPs for drug delivery has decreased significantly in recent years

6.1.2 Properties of Magnetic Nanoparticles

From many years magnets have been used for industrial and engineering applications. In recent years magnetic nanoparticles (MNPs) are used as drug carriers for human therapy. Magnetic field can make MNPs intelligent particles that can be guided remotely by changing the magnetic gradient and intensity. The magnetic properties of these materials are the result of magnetic moment of electrons. Each electron in the atom has a magnetic moment, which arises from two sources: one related to the orbital motion of electrons around the nucleus and another related to the rotation around the axis that is called spine. Thus each electron in an atom with orbital and spin moment can permanently act like a tiny magnet. Magnetization is a base of classification for magnetic materials. Accordingly, the materials are classified into three groups: ferromagnetic, paramagnetic, and diamagnetic. Magnetization of diamagnetic materials is negative and very low. The resultant magnetic dipole moment in these materials is zero. All gases (except oxygen), water, silver, gold, copper, diamond, graphite, bismuth, and many organic compounds are diamagnetic materials. Induced magnetic dipole in these materials is opposite of the external magnetic field and with elimination of external field the induced magnetic dipole is cancelled.

In paramagnetic materials bipolar magnetic orientation is not regular. As a result, these materials are not magnetic. If the material is placed in a magnetic field, the magnetic field lines orient quickly to the external magnetic field. By removing the magnetic field, the magnetic bipolar quickly returns to the previous status. Magnetization of these materials is positive and about 10^{-6} to 10^{-1}. Manganese, platinum, aluminum, and alkali metals are paramagnetic materials.

Ferromagnetic materials are like paramagnetic materials but they have the set of dipoles with same direction. MNPs used in medical applications belong to the paramagnetic materials. Also, particle coating and surfactant are the main factors in MNPs. A proper coating can prevent nanoparticles from being oxidized and swallowed by white cells. Polysaccharide coatings like dextran and chitosan are widely used in biological field. Hydrophilic surface coating results in a stable and uniform distribution of particles in the water and prevents unwanted ties and digestion of cells. The residence time of particles in the body depends on particle size, particle charge, hydrophilic surface, and intensity of magnetic field. These parameters are related to each other and the effect of each of them is a challenge in research [3].

Novel physical, mechanical, and chemical properties can be observed when the size of materials approaches the nanoscale. The physical properties include optical, magnetic, and electrical properties. It was found that magnetic nanoparticles behave differently from larger ones especially in magnetization. One of the most important properties that should be considered in biomedical application is residual magnetism in materials. It should be canceled when the external magnetic field is off. Special restriction, such as possible reaction, is required for MNPs materials that are used in biological field.

6.2 GOVERNING EQUATIONS

The basic equations in fluid mechanics and heat transfer are used to study fluid flow and heat transfer in drug delivery applications. The general equations are manipulated to present different case studies. They are used to solve these specific problems and calculate the important and effective parameters needed for drug delivery. For non–Newtonian fluids with constant properties the governing two-dimensional biofluid equations in Cartesian coordinate are as follows:

Continuity equation:

$$\frac{\partial u}{\partial x} + \frac{\partial v}{\partial y} = 0 \tag{6.1}$$

X-component of momentum equation:

$$\rho\left(u\frac{\partial u}{\partial x} + v\frac{\partial u}{\partial y}\right) = -\frac{\partial P}{\partial x} + \mu\left(\frac{\partial^2 u}{\partial x^2} + \frac{\partial^2 u}{\partial y^2}\right) + F_x \tag{6.2}$$

Y-component of momentum equation:

$$\rho\left(u\frac{\partial v}{\partial x} + v\frac{\partial v}{\partial y}\right) = -\frac{\partial P}{\partial y} + \mu\left(\frac{\partial^2 v}{\partial x^2} + \frac{\partial^2 v}{\partial y^2}\right) + F_y \tag{6.3}$$

where u, v, P, ρ, μ are velocity components in x and y directions, pressure, density, and viscosity, respectively. F_x is the horizontal component and F_y is the vertical component of magnetic body force that is applied to blood.

Energy equation:

$$\rho C_p\left(\frac{\partial T}{\partial t} + u\frac{\partial T}{\partial x} + v\frac{\partial T}{\partial y}\right) = k\left(\frac{\partial^2 T}{\partial x^2} + \frac{\partial^2 T}{\partial y^2}\right)$$
$$+ \mu\left\{2\left(\frac{\partial u}{\partial x}\right)^2 + 2\left(\frac{\partial v}{\partial x}\right)^2 + \left(\frac{\partial v}{\partial x} + \frac{\partial u}{\partial y}\right)^2\right\} + S \tag{6.4}$$

where T, C_p, k, and S are temperature, specific heat, thermal conductivity, and source term, respectively.

One important way to move the drug carriers through the body and deliver them to the designated place is the use of source term especially external source terms. Among various external source terms, magnetic field, electromagnetic field, and ultrasonic wave are most widely utilized tools for delivering the loaded drug to the designated place. The following section discusses different source terms and presents their general equations.

6.2.1 Magnetohydrodynamics Source Term

The magnetofluid dynamics or hydromagnetics is the study of the magnetic properties of electrically conducting fluids. This field of magnetohydrodynamics (MHD) was first initiated by Hannes Alfvén [4], for which he received the Nobel Prize in Physics in 1970. The fundamental concept behind MHD is that magnetic fields can induce currents in a moving conductive fluid, which in turn polarize the fluid and reciprocally change the magnetic field itself. The Navier–Stokes equations of fluid mechanics and the Maxwell's equations of electromagnetism describe the MHD phenomena. The effect of Maxwell's equations are shown as the MHD Force (source term) in the Navier–Stokes equations. This is the force created by the magnetic field when applied to an electrically conducting fluid such as blood and is given by Lorentz Force $\left(\vec{F}\right)$ as follows:

$$\vec{F} = \vec{J} \times \vec{B} \tag{6.5}$$

where $\vec{J}$, $\vec{E}$, and $\vec{B}$ are the current density, the electric field intensity vector, and the magnetic density vector. The current density is calculated from Ohm's law as below:

$$\vec{J} = \vec{E} + \sigma\left(\vec{u} \times \vec{B}\right) \tag{6.6}$$

where $\vec{u}$ and σ are the flow velocity and the electrical conductivity of the fluid.

The source term due to MHD effect in energy equation is given by [5]:

$$S = \frac{J^2}{\sigma} = \sigma\left(E^2 + u^2B - 2uB\right) \tag{6.7}$$

6.2.2 Electromagnetic Source Term

Electromagnetism is a branch of physics that involves the study of electromagnetic force, a type of physical interaction that occurs between electrically charged particles. The electromagnetic force usually shows electromagnetic fields, such as electric fields, magnetic fields, and light. The electromagnetic force is one of the four fundamental interactions in the nature.

Electromagnetic force is the force created by magnetic field when applied to fluids such as blood and is expressed by the Maxwell equations as [6]:

$$\nabla \cdot E = \frac{\rho}{\mu_0} \tag{6.8}$$

$$\nabla \cdot B = 0 \tag{6.9}$$

$$\nabla \times E = \frac{\partial B}{\partial t} \tag{6.10}$$

$$\nabla \times B = \varepsilon_0 \left(J + \mu_0 \frac{\partial E}{\partial t} \right) \qquad (6.11)$$

where E is the electrical field, μ_0 is the permeability of a vacuum, and ε_0 is the permeability of free space and is equal to $\varepsilon_0 = 4\pi/C$, where C is the speed of the light, and J is the vector current density. For steady condition the magnetic induction, B, in the external magnetic field is given by:

$$B = \mu_0(M + H) \qquad (6.12)$$

where μ_0 is the permeability of free space. Magnetization (M) is given by:

$$M = \chi H \qquad (6.13)$$

where χ is dimensionless volumetric magnetic susceptibility and $\vec{H}$ is external magnetic field strength vector.

By applying magnetic field as a body force in drug delivery, the magnetic volume body force in the x and y directions (F_x and F_y) as well as the energy source term are determined by, respectively [7]:

$$F_x = \left[\frac{\mu_0}{2} \chi \frac{\partial}{\partial x} (\vec{H} \cdot \vec{H}) \right] C_V \qquad (6.14)$$

$$F_y = \left[\frac{\mu_0}{2} \chi \frac{\partial}{\partial y} (\vec{H} \cdot \vec{H}) \right] C_V \qquad (6.15)$$

$$S_{energy} = \mu_0 T \frac{\partial M}{\partial T} \left[u \frac{\partial H}{\partial x} + v \frac{\partial H}{\partial y} \right] \qquad (6.16)$$

Each MNP in biofluid (blood) flow is absorbed by external magnetic field and is modeled by [8]:

$$\vec{F}_{Mag} = 0.5(\text{Volume})_P \mu_0 \chi \vec{\nabla} |H^2| \qquad (6.17)$$

where H, μ_0, and χ are external magnetic field, magnetic permeability of vacuum, and magnetic susceptibility of the particles, respectively. Subscript P refers to particle. For n particles magnetic force is equal to:

$$F_x = \left[\frac{1}{2} \mu_0 \chi \frac{\partial}{\partial x} (\vec{H} \cdot \vec{H}) \right] C_V \qquad (6.18)$$

$$F_y = \left[\frac{1}{2} \mu_0 \chi \frac{\partial}{\partial y} (\vec{H} \cdot \vec{H}) \right] C_V \qquad (6.19)$$

where F_X and F_y are the effect of particle aggregation near the magnet on the blood flow.

$$F_x = n_P F_{Mag-x} \tag{6.20}$$

$$F_y = n_P F_{Mag-y} \tag{6.21}$$

Volume concentration, C_v, is defined as:

$$C_V = n_P (\text{Volume})_P \tag{6.22}$$

Nondimensional concentration, C, is defined as:

$$C = \frac{C_V}{C_0} \tag{6.23}$$

where C_0 is the initial concentration and is set before the beginning of the test or the numerical simulation.

Particles concentration is calculated from:

$$\frac{\partial C}{\partial t} + \vec{\nabla} \cdot (C\vec{v}_p) = \vec{\nabla} \cdot (D\vec{\nabla}C) \tag{6.24}$$

where V_P is particle velocity and is calculated by applying the Newton equation to each particle for uniform motion as:

$$\vec{F}_{Drag} = 6\pi\mu D_P(\vec{v_P} - \vec{v_f}) = \vec{F}_{Mag} \rightarrow \vec{v_P} = \frac{\vec{F}_{Mag}}{6\pi\mu D_P} + \vec{v_f} \tag{6.25}$$

$$\vec{v_f} = u\hat{i} + v\hat{j} \tag{6.26}$$

where V_f is blood flow velocity and D_P is particle diameter. Diffusion coefficient, D, is calculated from Einstein equation [7]:

$$D = \frac{k_B T}{6\pi\mu D_P} \tag{6.27}$$

where k_B is the Boltzmann constant and T is the absolute temperature.

6.2.3 Acoustic Radiation Force

Acoustic radiation force produces a steady flow that also is known as acoustic streaming (AS). Acoustic radiation forces occur when High-Intensity Focused Ultrasound is utilized. The acoustic force is able to direct and displace the particles that come in contact with in the direction of applied ultrasonic wave. The particles absorb or reflect the traveling ultrasound wave and create a momentum called primary ultrasound radiation force (USRF). The USRF causes the particle to displace. Radiation forces are proportional to the absorption coefficient of the tissue and the applied energy.

The acoustic radiation forces are commonly used to displace microbubbles toward the vessel wall. The microbubble movements induce shear forces and gaps in the vessel wall endothelium and cause uptake in that area [9]. In addition to the primary USRF, a secondary USRF also acts between individual bubbles to increase the attraction between each other, causing a greater concentration at the target.

To determine the pressure, velocity, and density the second-order perturbation theory is applied as follows, respectively:

$$
\begin{aligned}
P &= P_0 + P_1 + P_2 \\
\vec{v} &= \vec{v}_0 + \vec{v}_1 + \vec{v}_2 \\
\rho(p) &= \rho_0 + \rho_1 + \rho_2
\end{aligned}
\tag{6.28}
$$

where P, v, and ρ are pressure, velocity, and density, respectively. The subscripts 0, 1, and 2 are related to zero-, first-, and second-order perturbation. The superscript T is used for transposed. μ and λ are viscosity and kinematic viscosity, respectively.

The zero-, first-, and second-order continuity and momentum equations are as follows, respectively:

$$
\frac{\partial \rho_0}{\partial t} = -\nabla \cdot (\rho_0 \vec{v}_0)
\tag{6.29}
$$

$$
\frac{\partial \rho_1}{\partial t} = -\nabla \cdot (\rho_0 \vec{v}_1 + \rho_1 \vec{v}_0)
\tag{6.30}
$$

$$
\frac{\partial \rho_2}{\partial t} = -\nabla \cdot (\rho_0 \vec{v}_2 + \rho_1 \vec{v}_1 + \rho_2 \vec{v}_0)
\tag{6.31}
$$

$$
\rho_0 \frac{\partial v_0}{\partial t} = -\nabla p_0 - \rho_0 (v_0 \cdot \nabla) v_0 + \nabla \cdot \left(\mu_0 \left(\nabla v_0 + \nabla v_0^T \right) \right) + \nabla (\lambda_0 (\nabla \cdot v_0))
\tag{6.32}
$$

$$
\begin{aligned}
\rho_0 \frac{\partial \vec{v}_1}{\partial t} + \rho_1 \frac{\partial \vec{v}_0}{\partial t} &= -\nabla p_1 - \rho_0 (\vec{v}_1 \cdot \nabla) \vec{v}_0 - \rho_0 (\vec{v}_0 \cdot \nabla) \vec{v}_1 - \rho_1 (\vec{v}_0 \cdot \nabla) \vec{v}_0 \\
&\quad + \nabla \cdot \left(\mu_0 \left(\nabla \vec{v}_1 + \nabla \vec{v}_1^T \right) \right) + \nabla \cdot \left(\mu_1 \left(\nabla \vec{v}_0 + \nabla \vec{v}_0^T \right) \right) \\
&\quad + \nabla \left(\lambda_0 (\nabla \cdot \vec{v}_1) \right)
\end{aligned}
\tag{6.33}
$$

$$
\begin{aligned}
\rho_0 \frac{\partial \vec{v}_2}{\partial t} + \rho_1 \frac{\partial \vec{v}_1}{\partial t} + \rho_2 \frac{\partial \vec{v}_0}{\partial t} &= -\nabla p_2 - \rho_0 (\vec{v}_0 \cdot \nabla) \vec{v}_2 - \rho_0 (\vec{v}_1 \cdot \nabla) \vec{v}_1 \\
&\quad - \rho_0 (\vec{v}_2 \cdot \nabla) \vec{v}_0 - \rho_1 (\vec{v}_0 \cdot \nabla) \vec{v}_1 - \rho_1 (\vec{v}_1 \cdot \nabla) \vec{v}_0 \\
&\quad - \rho_2 (\vec{v}_0 \cdot \nabla) \vec{v}_0 + \nabla \cdot \left(\mu_0 \left(\nabla \vec{v}_2 + \nabla \vec{v}_2^T \right) \right) \\
&\quad + \nabla \cdot \left(\mu_1 \left(\nabla \vec{v}_1 + \nabla \vec{v}_1^T \right) \right) + \nabla \left(\lambda_0 (\nabla \cdot \vec{v}_2) \right)
\end{aligned}
\tag{6.34}
$$

6.3 APPLICATION OF MAGNETIC NANOPARTICLES IN DRUG DELIVERY

As mentioned earlier, magnetic field is one of the most useful external forces for moving the nanodrug carrier through the body and delivering the drug to the desired place. One of the drug carrier options is MNPs. MNPs are typically made from iron, nickel, and cobalt. They are loaded with drugs, coated with biocompatible coatings, and injected into the human blood vessel (blood stream). The loaded particles move through the vessel and are absorbed by the cancerous solid tumor where the magnetic field is applied.

The effect of external magnetic field on the MNP concentration in cancerous solid tumor is presented later in the discussion. In this model the MNPs diffusion coefficients in the capillary as well as its wall and the tumor tissue are considered as variable and calculated. These coefficients are functions of MNP diameter, pore size of capillary wall, and tissue porosity.

Fig. 6.1 depicts the schematic model of the capillary, its wall (the endothelial layer), the tumor tissue, and external magnet. The external magnet is cylindrical with 4 mm diameter and is placed in the middle of the space on the top of the tumor tissue.

The MNPs have a core—shell shape (superparamagnetic metallic core with 5-nm biocompatible shell). The blood is treated as non-Newtonian, its density is 1050 kg/m^3, and its viscosity is calculated by the power law as follows [10]:

$$\mu = m\dot{\gamma}^{n-1} \tag{6.35}$$

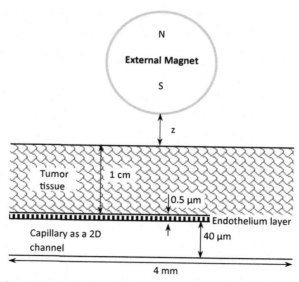

Figure 6.1 Schematic model of capillary, its wall, tumor tissue, and external magnet.

where $\dot{\gamma}$ is the blood shear rate and m and n are constants and equal to 0.012 and 0.8, respectively [10].

The magnetic body force inside the capillary is given by [11]:

$$\overrightarrow{\mathbf{F}} = \overrightarrow{\mathbf{F}}_1 n_p = 0.5 \times \forall_{core} \times \mu_0 \frac{\chi_{MNP}}{1 + \frac{\chi_{MNP}}{3}} \times \nabla |H|^2 \times n_p \quad \text{and}$$

$$n_p = \frac{C_{MNP}}{\forall_{MNP}} = C \times \frac{C_0}{\forall_{MNP}}$$

(6.36)

where $\forall_{MNP}$ is a single MNP's volume and χ_{MNP} is the magnetic susceptibility of the MNPs and set equal to 3 [12]. C_{MNP}, C_0, and C are volumetric concentration of MNPs in the blood, concentration of MNPs at inlet, and dimensionless concentration (C_{MNP}/C_0), respectively.

Inserting Eq. (6.36) in Eqs. (6.1)–(6.3), the horizontal and vertical momentum equations become, respectively:

$$\rho \left(u \frac{\partial u}{\partial x} + v \frac{\partial u}{\partial y} \right) = -\frac{\partial P}{\partial x} + \mu \left(\frac{\partial^2 u}{\partial x^2} + \frac{\partial^2 u}{\partial y^2} \right) + \left[\frac{1}{2}\mu_0 \frac{\chi_{MNP}}{1 + \frac{\chi_{MNP}}{3}} \frac{\forall_{core}}{\forall_{MNP}} \frac{\partial}{\partial x}|H|^2 \right] C_0 C$$

(6.37)

$$\rho \left(u \frac{\partial v}{\partial x} + v \frac{\partial v}{\partial y} \right) = -\frac{\partial P}{\partial y} + \mu \left(\frac{\partial^2 v}{\partial x^2} + \frac{\partial^2 v}{\partial y^2} \right) + \left[\frac{1}{2}\mu_0 \frac{\chi_{MNP}}{1 + \frac{\chi_{MNP}}{3}} \frac{\forall_{core}}{\forall_{MNP}} \frac{\partial}{\partial y}|H|^2 \right] C_0 C$$

(6.38)

The concentration equation inside the capillary is as follows [10,13,14]:

$$\frac{\partial C}{\partial t} + \nabla \cdot (C \overrightarrow{\mathbf{v}}_{MNP}) = \nabla \cdot (D_{blood} \nabla C)$$

(6.39)

where D_{blood}, $\overrightarrow{\mathbf{v}}_{MNP}$, and $V_{relative}$ are MNPs diffusion coefficient in blood, MNPs velocity vector, and MNPs relative velocity with respect to the blood flow, respectively [13,14].

$$\overrightarrow{\mathbf{v}}_{MNP} = \overrightarrow{\mathbf{v}} + \overrightarrow{\mathbf{v}}_{relative}, \overrightarrow{\mathbf{v}}_{relative} = \frac{\overrightarrow{\mathbf{F}}_1}{6\pi \mu_{blood} r_{MNP}}$$

(6.40)

$$D_{Blood} = D_B + D_S, D_B = \frac{k_B T}{6\pi \mu_{blood} r_{MNP}}$$

(6.41)

where D_B is Brownian diffusion coefficient and is calculated from Einstein relation [13] and D_s is scattering diffusion coefficient and equal to 3.5×10^{-12} [15].

$$\frac{\partial C}{\partial t} = -\nabla \cdot \left[-D_{Blood} \nabla C + C \overrightarrow{v} + C \frac{0.5 \forall_{core}}{6\pi \mu r_{MNP}} \mu_0 \frac{\chi_{MNP}}{1 + \frac{\chi_{MNP}}{3}} \nabla \left(|H|^2 \right) \right]$$

(6.42)

The capillary wall (endothelial layer) is modeled as a saturated porous medium and the concentration equation is calculated by [16]:

$$\frac{\partial C}{\partial t} = -\nabla \cdot [-D_{Endo}\nabla C + C\vec{\mathbf{v}}_{MNP}] + \frac{G}{\varepsilon} - E(C) \tag{6.43}$$

where G and E are generation and uptake terms, respectively. In this model no generation and no uptake are considered.

The MNPs velocity ($\vec{\mathbf{v}}_{MNP}$) is the same as relative velocity as follows [17]:

$$\vec{\mathbf{v}}_{relative} = \left(\frac{F}{\lambda_g^2}\right) \frac{\vec{\mathbf{F}}_1}{6\pi\mu r_{MNP}} \tag{6.44}$$

The concentration equation for endothelium is obtained by [15]:

$$\frac{\partial C}{\partial t} = -\nabla \cdot \left[-D_{Endo}\nabla C + C\left(\frac{F}{\lambda_g^2}\right) \frac{0.5\mathbf{V}_{core}}{6\pi\mu_{plasma}r_{MNP}} \mu_0 \frac{\chi_{MNP}}{1 + \frac{\chi_{MNP}}{3}} \nabla(|H|^2) \right] \tag{6.45}$$

The first term in the right-hand side of Eq. (6.45) is penetration due to concentration gradient and the second term is penetration under the influence of magnet (magnetic term).

Endothelium diffusion coefficient (D_{Endo}) is given by [16,18]:

$$D_{endo} = D_{\infty} \times \left(\frac{\varepsilon}{\lambda_g^2}\right) \times S \times J \tag{6.46}$$

where D_{∞} is diffusion coefficient of particle in unbounded fluid. S (steric coefficient), J (hydrodynamic coefficient), and ε (the porosity of endothelial layer) are calculated as follows [18,19]:

$$S = (1 - \alpha)^2, \quad \alpha = \frac{r_{MNP}}{r_{pore}} \tag{6.47}$$

$$J = \left(1 - 2.1044\alpha + 2.089\alpha^3 - 0.948\alpha^5\right) \tag{6.48}$$

$$\varepsilon = \frac{\text{Intercellular gap in Endotheium layer}}{\text{Average size of Endotheium cell}} \tag{6.49}$$

By incorporating Eqs. (6.47)–(6.49) into Eq. (6.46) the diffusion coefficient in endothelial layer becomes as:

$$D_{endo} = D_{\infty} \times \left(\frac{\varepsilon}{\lambda_g^2}\right)(1 - \alpha)^2\left(1 - 2.1044\alpha + 2.089\alpha^3 - 0.948\alpha^5\right),$$

$$\alpha = \frac{2 \times r_{MNP}}{\text{Intercellular gap}} \tag{6.50}$$

The gaps between endothelial cells are filled with plasma. The diffusion coefficient of particle in unbounded fluid, D_∞, is given by Brownian diffusion coefficient of particles in the plasma as follows:

$$D_{plasma} = \frac{k_B T}{6\pi\mu_{plasma}r_{MNP}} \tag{6.51}$$

where μ_{plasma} is plasma viscosity and it is equal to 1.24 mPa s [20,21]. The concentration equation for tumor tissue, a porous media, is:

$$\frac{\partial C}{\partial t} = -\nabla \cdot \left[-D_{Tissue}\nabla C + C\left(\frac{F}{\lambda_g^2}\right)\frac{0.5\forall_{core}}{6\pi\mu_{plasma}r_{MNP}}\mu_0\frac{\chi_{MNP}}{1+\frac{\chi_{MNP}}{3}}\nabla\left(|H|^2\right) \right] \tag{6.52}$$

where D_{Tissue} is the MNPs diffusion coefficient in the tissue and is obtained by:

$$D_{Tissue} = D_\infty \times \left(\frac{\varepsilon}{\lambda_g^2}\right) \times S \times J \tag{6.53}$$

Because the interstitial fluid is motionless D_∞ is usually equal to D_{plasma} [17]. The steric coefficient (S) and hydrodynamic coefficient (J) are obtained from [22,23]:

$$S = \exp\left(-0.84k^{1.09}\right), \quad k = \left(1 + \frac{r_{MNP}}{r_{fiber}}\right)^2 \times \phi \tag{6.54}$$

$$J = e^{-\alpha \times \phi^\vartheta} \tag{6.55}$$

where ϕ is the fibers volume fraction. Collagen fibril radius, $r_{fibrils}$, is equal to 50 nm. The value of $r_{fibrils}$ is between 15 and 100 nm [24,25]. Also, the value of the fiber volume fraction of fibers, ϕ is equal to 0.01 where the values for three different tumors are $\phi = 0.01$ (LS174T), $\phi = 0.03$ (HSTS26T), and $\phi = 0.045$ (U87) [26].

The geometrical tortuosity (λ_g) is expressed as a function of porosity [16]:

$$\lambda_g = \varepsilon^{-n} \tag{6.56}$$

Tissue porosity, ε, varies between 0.06 and 0.6 for different tumors [27–29]. The value of n has an upper and lower limit and is determined by:

$$\text{Upper limit}: \quad n = 0.23 + 0.3\,\varepsilon + \varepsilon^2 \tag{6.57}$$

$$\text{Lower limit}: \quad n = 0.23 + \varepsilon^2 \tag{6.58}$$

For safe simplicity an average value of n is utilized.

Table 6.2 Boundary conditions for Fig. 6.1

Boundary	Conditions
Channel inlet	Steady velocity ($U_{in} = 0.2$ mm/s [10])
	Steady magnetic nanoparticles (MNPs) inlet concentration ($C_0 = 10^{-4}$)
Channel upper wall	No-slip condition ($u,v = 0$)
	MNPs can go through the wall
Channel lower wall	No-slip condition ($u,v = 0$)
	MNPs cannot go through the wall
Channel outlet	Neumann boundary condition for both momentum and concentration equations
Tissue upper, right, and left wall	MNPs cannot go through the wall

According to Fig. 6.1, the boundary conditions are as provided in Table 6.2.

For initial condition dimensionless MNPs concentration (C) inside the vessel is equal to 1 and no concentration is assumed in the capillary wall (endothelial layer) and in the tumor tissue ($C = 0$).

The calculated MNPs diffusion coefficients in the endothelial layer and in the tumor tissue ($\varepsilon = 0.3$) are shown in Fig. 6.2.

As shown, the diffusion coefficient in the endothelial layer decreases as MNPs diameter increases. The tumor tissue diffusion coefficient is maximized at 100 nm and decreases as the MNPs diameter increases. As shown, the endothelial layer diffusion coefficient is more sensitive to MNPs size than the tumor tissue.

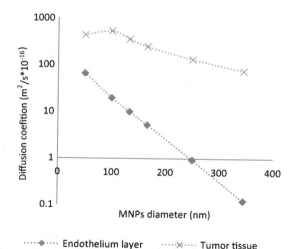

Figure 6.2 The magnetic nanoparticles (MNPs) diffusion coefficients in endothelial layer and the tumor tissue (porosity = 0.3).

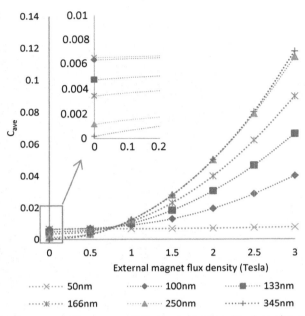

Figure 6.3 Dimensionless magnetic nanoparticle concentration (C_{ave}) in the presence of external magnet with different flux density (porosity = 0.3, $z = 10$ mm).

The average dimensionless MNPs concentration parameter in the tissue (C_{ave}) is [22]:

$$C_{ave} = \frac{\sum C_i A_i}{\sum A_i} \qquad (6.59)$$

where C_i is the MNPs dimensionless concentration, A_i is the area of i-th computational cell and are summed over the whole tissue. The dimensionless MNPs concentration (C_{ave}) in the presence of external magnet is shown in Fig. 6.3.

The average dimensionless MNPs concentration (C_{ave}) within the tumor increases as MNPs diameter increases from 50 to 250 nm. But average dimensionless concentration of 250 and 345 nm MNPs overlap each other. This is because as MNPs size increases from 250 to 345 nm, the hydraulic and steric coefficients of MNPs in the endothelial layer decrease. Furthermore, not only the concentration of MNPs greater than 250 nm within the entire tumor tissue is high but also MNPs (drug careers) reach into the whole tumor tissue as well. This observation is useful for medical treatments as it increases the treatment effectiveness due to delivering the drugs to whole tumors and also minimizes the drug side effects because those MNPs do not pass the endothelial layer (capillary wall) in most healthy organs. The dimensionless MNPs concentration within the capillary and the tissue ($x = 1/2$) at 24 h time in the absence and the presence of 2 T external magnet is represented in Figs. 6.4 and 6.5, respectively.

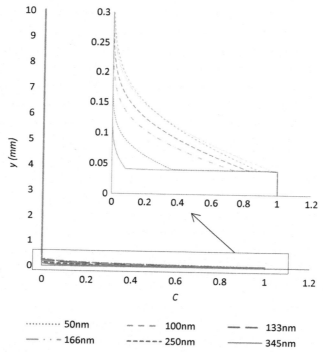

Figure 6.4 Magnetic nanoparticle dimensionless concentration inside the capillary as well as tissue ($x = l/2$) in the absence of external magnet (porosity = 0.3, $z = 10$ mm).

As shown, the 250- and 345-nm MNPs agglomerate on the upper capillary wall. This is because magnetic forces acting on the large MNPs are strong while their steric and hydraulic coefficients in the endothelial layer are low. Therefore the MNPs cannot easily pass through the endothelial layer and thus, they agglomerate on the upper wall of capillary.

The average dimensionless concentration (C_{ave}) for different external magnet distances from the tumor tissue, z, is depicted in Fig. 6.6.

As shown, for the MNPs larger than 50 nm, the average dimensionless concentration increases exponentially by decreasing the distance between the external magnet and tumor tissue. This is because the magnetic force acting upon the MNPs increases as the distance of the external magnet from the tumor tissue decreases.

The average dimensionless MNPs concentration (C_{ave}) within the tumor increases as MNPs diameter increases from 50 to 250 nm. But average dimensionless concentrations of 250- and 345-nm MNPs overlap each other. This is because as MNPs size increases from 250 to 345 nm, the hydraulic and steric coefficients of the endothelial layer decrease. Also, MNPs greater than 250 nm cannot easily pass through the endothelial layer due to low hydraulic and steric coefficients and thus, they agglomerate on the upper

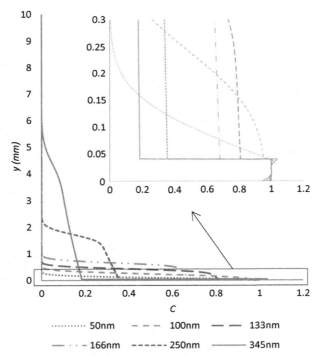

Figure 6.5 Magnetic nanoparticle dimensionless concentration within the capillary and the tissue ($x = l/2$) under the influence of 2 T external magnet (porosity $= 0.3$, $z = 10$ mm).

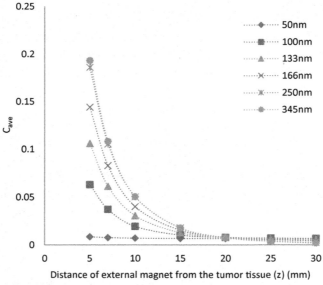

Figure 6.6 Dimensionless magnetic nanoparticle concentration (C_{ave}) in the presence of external magnet for different external magnet distances from the tumor (porosity $= 0.3$, $B_0 = 2$ T).

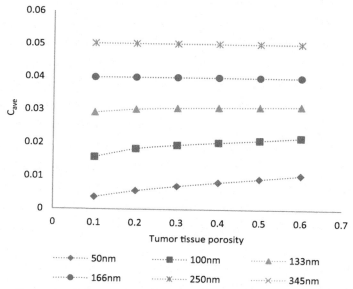

Figure 6.7 Magnetic nanoparticle average dimensionless concentration within the tumor tissue with different porosity ($z = 10$ mm, $B = 2$ T).

wall of capillary. The average dimensionless concentration (C_{ave}) in the tumor tissue with different porosities (ε) is represented in Fig. 6.7.

As shown, the porosity does not affect the average dimensionless concentration for MNPs greater than 100 nm. This is because, as mentioned, the concentration of large MNPs is limited by the rate of passing through the endothelial layer.

Fig. 6.8 represents the average dimensionless MNPs concentration (C_{ave}) with and without external magnet.

As shown, in the absence of external magnet, the average dimensionless concentration decreases as time passes. This is because the gradient of MNPs concentration that causes the diffusion decreases as time passes. Also, concentration variation of MNPs greater than 100 nm is linear with time in the presence of external magnet. This is because the magnetic term of MNPs penetration in the tissue does not vary with time.

6.4 APPLICATION OF ACOUSTIC STREAMING IN DRUG DELIVERY

Ultrasonic waves have a variety of applications in bio field. The most important applications are diagnosis and treatment of diseases, drug delivery, cell separation, and cell study. Ultrasonic wave application in drug delivery reduces the harmful effects and increases the penetration of the drug into the desired unit. In this case study, the interaction of two nonlinear phenomena, AS due to passing ultrasonic waves through biofluid

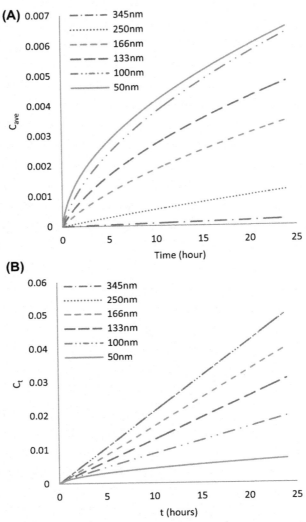

Figure 6.8 Average dimensionless magnetic nanoparticle concentration (C_{ave}) variation with time: (A) in the absence of external magnet; (B) in the presence of 2 T external magnet (porosity = 0.3, z = 10 mm).

and non-Newtonian viscosity is presented. Taking into account the nonlinear effects of ultrasonic field, continuity, and momentum and state equations are coupled and solved. The parametric effects of inlet flow velocity and non-Newtonian viscosity models on AS are investigated.

Schematic of the geometry is depicted in Fig. 6.9.

As shown, the channel is assumed to be two dimensional, 6 mm by 1 mm, which resembles the coronary artery in the human body. Blood flows into the vessel with a mean velocity of 6 mm/s. The bottom wall of the channel vibrates due to ultrasonic

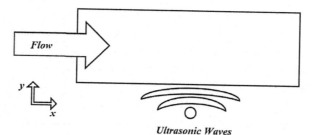

Figure 6.9 Schematic of channel that vibrates due to ultrasonic wave.

wave propagation. Therefore it is modeled by corresponding velocity of vibration. It is assumed that a half wavelength standing wave of 750 KHz frequency is developed. As a result the top channel wall vibrates as well.

To solve this problem, in addition to Eqs. (6.28)–(6.34), the following equations are used.

The second-order of state equation is:

$$p_2 = c^2 \rho_2 + \frac{\partial c^2}{\partial \rho} \rho_1^2 \tag{6.60}$$

The relation between density and pressure is provided by:

$$\rho_1 = \rho_0 k p_1 \tag{6.61}$$

where c and k are speed of sound and bulk module.

Due to velocity variation the viscosity is perturbed and the first-order viscosity change is as follows:

$$\mu(\dot{\gamma}) = \mu_0 + \mu_1 \tag{6.62}$$

Blood is assumed to be Newtonian and the Casson viscosity model is used:

$$\mu_0 = \left(\sqrt{\mu_c} + \sqrt{\frac{\tau_c}{\dot{\gamma}}} \right)^2 \tag{6.63}$$

where μ_c and τ_c are constant. The shear stress rate ($\dot{\gamma}$), symmetric matrix (D), and viscosity (μ_1) are calculated by the following equations, respectively:

$$\dot{\gamma} = \sqrt{2\left((trD)^2 - trD^2\right)} \tag{6.64}$$

$$D_{ij} = \frac{1}{2}\left(\frac{\partial u_i}{\partial x_j} + \frac{\partial u_j}{\partial x_i} \right) \tag{6.65}$$

Table 6.3 Boundary conditions for Fig. 6.9

Boundary	Condition
Vertical boundary (left side)	$V = 1$, 2, or 3 mm/s
Vertical boundary (right side)	P is the same as left side
Horizontal boundary	No-slip condition
For solving the first-order perturbation horizontal boundaries	Perturbation velocities are set to have standing waves in channel, for example, for a half-length standing wave, the magnitude of a wall velocity is minus of the other
For solving the second-order perturbation horizontal boundaries	Velocity magnitude and phase are determined from continuity equation
For solving first- and second-order perturbation vertical boundaries	Normal impedance is set to model a soft, hard, or lossy wall. Soft wall has zero impedance, whereas impedance of hard wall is infinity and lossy wall impedance is a positive value

$$\mu_1 = \left(\frac{\partial \mu}{\partial \dot{\gamma}}\right)_0 \dot{\gamma}_1 \tag{6.66}$$

According to Fig. 6.9, the boundary conditions are as provided in Table 6.3.

The AS velocity of three different viscosity models for soft-wall boundary condition is depicted in Fig. 6.10.

As shown the velocity behavior changes with different non-Newtonian viscosity models at the beginning and at the end of the channel.

Fig. 6.11 shows the effect of the three different viscosity models on the AS velocity for lossy wall.

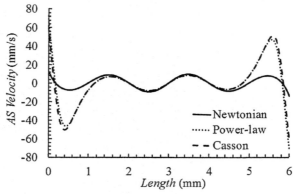

Figure 6.10 Effect of different viscosity models on the acoustic streaming (AS) velocity for soft-wall condition.

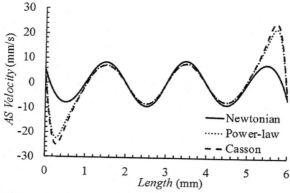

Figure 6.11 Effect of different viscosity models on the acoustic streaming (AS) velocity for lossy-wall condition.

As shown the velocity behaves as in the previous case, Fig. 6.10. However, the magnitude of velocity decreases at the pick point.

The effect of different viscosity models on the (AS) velocity for hard–wall condition is depicted in Fig. 6.12.

As shown, little change is observed for three viscosity models.

The effect of inlet velocity on first-order velocity is depicted in Fig. 6.13.

As shown its effect is minimal and negligible and the results are the same.

The second–order velocity (AS) is affected by the inlet velocity, as shown by Fig. 6.14.

By changing the inlet velocity from 1 to 3 mm/s, the velocity magnitude is increased to 4 mm/s, see Fig. 6.14. This means that increasing the inlet velocity increased the AS resistance due to increasing the velocity gradient.

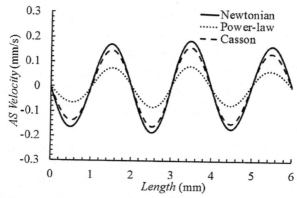

Figure 6.12 Effect of different viscosity models on the acoustic streaming (AS) velocity for hard-wall condition.

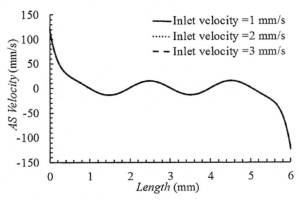

Figure 6.13 Effect of the inlet velocity magnitude on the first-order velocity.

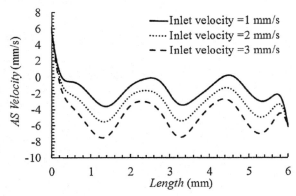

Figure 6.14 Effect of the inlet velocity magnitude on the second-order (AS) velocity.

REFERENCES

[1] N. Bertrand, J.C. Leroux, The journey of a drug-carrier in the body: an anatomo-physiological perspective, Journal of Controlled Release 161 (2012) 152−163.
[2] S. Bamrungsap, et al., Nanotechnology in therapeutics: a focus on nanoparticles as a drug delivery system, Nanomedicine 7 (8) (2012) 1253−1271.
[3] O.D.E. Peppas, Opsonization, Biodistribution,and Pharmacokinetics of polymeric nanoparticles, International Journal of Pharmaceutics 307 (2006) 93−102.
[4] P.A. Davidson, An Introduction to Magnetohydrodynamics, University of Cambridge, 2001.
[5] A. Shahidian, M. Ghassemi, Effect of magnetic flux density and other properties on temperature and velocity distribution in magnetohydrodynamic (MHD) pump, IEEE Transactions on Magnetics 45 (1) (January 2009).
[6] R. Fitzpatrick, Maxwell's Equations and the Principles of Electromagnetism, Jones & Bartlett Publishers, 2008.
[7] J. Berthier, P. Silberzan, Microfluidics for Biotechnology, 2006. Available from: http://onlinelibrary.wiley.com/doi/10.1002/cbdv.200490137/abstract.

[8]   D. Hautot, et al., Preliminary evaluation of nanoscale biogenic magnetite in Alzheimer's disease brain tissue, Proceedings Biological Sciences 270 (Suppl. 1) (2003) S62–S64.

[9]   A.F.H. Lum, et al., Ultrasound radiation force enables targeted deposition of model drug carriers loaded on microbubbles, Journal of Controlled Release 111 (1–2) (2006) 128–134.

[10] M.R. Habibi, M. Ghasemi, Numerical study of magnetic nanoparticles concentration in biofluid (blood) under influence of high gradient magnetic field, Journal of Magnetism and Magnetic Materials 323 (2011) 32–38.

[11] M.R. Mohammadi, Analysis of Blood Flow and Heat Transfer under the Effect of Magnetic Field, Master of science thesis, K.N. Toosi University of Technology, 2014.

[12] T. Lunnoo, T. Puangmali, Capture efficiency of biocompatible magnetic nanoparticles in arterial flow: a computer simulation for magnetic drug targeting, Nanoscale Research Letters 10 (2015) 426.

[13] A. Grief, G. Richardson, Mathematical modelling of magnetically targeted drug delivery, Journal of Magnetism and Magnetic Materials 293 (2005) 455–463.

[14] M.R. Habibi, M. Ghassemi, M.H. Hamedi, Analysis of high gradient magnetic field effects on distribution of nanoparticles injected into pulsatile blood stream, Journal of Magnetism and Magnetic Materials 324 (2012) 1473–1482.

[15] A. Nacev, C. Beni, O. Bruno, B. Shapiro, The behaviors of ferromagnetic nano-particles in and around blood vessels under applied magnetic fields, Journal of Magnetism and Magnetic Materials 323 (2011) 651–668.

[16] k. Khanafer, k. Vafai, The role of porous media in biomedical engineering as related to magnetic resonance imaging and drug delivery, Heat Mass Transfer 42 (2006) 939–953.

[17] R.K. Jain, L.T. Baxter, Mechanisms of heterogeneous distribution of monoclonal antibodies and other macromolecules in tumors: significance of elevated interstitial pressure, Cancer Research 48 (1988) 7022–7032.

[18] R.L. Fournier, Basic Transport Phenomena in Biomedical Engineering, CRC Press, New York, 2011.

[19] L. Ai, K. Vafai, A coupling model for macromolecule transport in a stenosed arterial wall, International Journal of Heat and Mass Transfer 49 (2006) 1568–1591.

[20] U. Windberger, A. Bartholovitsch, R. Plasenzotti, K.J. Korak, G. Heinze, Whole blood viscosity, plasma viscosity and erythrocyte aggregation in nine mammalian species reference values and comparison of data, Experimental Physiology: Translation and Integration 88 (3) (2003) 431–440.

[21] G. Késmárky, Kenyeres, M. Rábai, K. Tóth, Plasma viscosity: a forgotten variable, Clinical Heortheology and Microcirculation 39 (2008) 243–246.

[22] W.M. Saltzman, Drug Delivery Engineering Principles for Drug Therapy, Oxford University Press, 2001.

[23] M.A. Swartz, M.E. Fleury, Interstitial flow and its effects in soft tissues, Annual Review of Biomedical Engineering 9 (2007) 229–256.

[24] M.P.E. Wenger, L. Bozec, M.A. Horton, P. Mesquida, Mechanical properties of collagen fibrils, Biophysical Journal 93 (2007) 1255–1263.

[25] D.H. Kim, et al., Nanoscale cues regulate the structure and function of macroscopic cardiac tissue constructs, Applied Biological Sciences 107 (2) (2010) 565–570.

[26] S. Ramanujan, et al., Diffusion and convection in collagen gels: implications for transport in the tumor interstitium, Biophysical Journal 83 (2002) 1650–1660.

[27] B.A. Graff, I. BjØrnæs, E.K. Rofstad, Macromolecule uptake in human melanoma xenografts: relationships to blood supply, vascular density, microvessel permeability and extracellular volume fraction, European Journal of Cancer 36 (2000) 1433–1440.

[28] Z. Bhujwalla, C. McCoy, J. Glickson, R. Gilfies, M. Stubbs, Estimations of intra- and extracellular volume and pH by 31P magnetic resonance spectroscopy: effect of therapy on RIF-1 tumours, British Journal of Cancer 78 (5) (1998) 606–611.

[29] R.K. Jain, Transport of molecules in the tumor interstitium: a review, Cancer Research 47 (1987) 3039–3051.

INDEX